KT-230-121

7000000002357

PATRICK MOORE

on the Moon

PATRICK MOORE
on the Moon

CASSELL&CO

Contents

Foreword

More than half a century ago, when the war was not long over, I was invited to write a book about the Moon. I did so; I called it *Guide to the Moon*. It was reprinted many times, but the last edition came out as long ago as 1977, and much has happened since then.

When I was invited to write this new book, only small parts of the old *Guide* could be retained, mainly those sections dealing with the Moon's movements. Therefore this book is new, though I have kept to the plan of concentrating upon the observational aspect.

Patrick Moore
Selsey, January 2001

1

The *Eagle* Has Landed

On the evening of 20 July 1969, I was in a BBC television studio, then at Lime Grove in Shepherd's Bush. It was a great occasion: Astronauts Neil Armstrong and Buzz Aldrin were on their way to the Moon. For the first time in human history, the gap between the two worlds was about to be bridged.

Of course the immediate concern was the safety of the astronauts, not forgetting Michael Collins, the third member of the Apollo team, who was patiently orbiting the Moon in the Command Module of the spacecraft. There had been serious suggestions that the waterless lunar 'seas', at least, might be covered with deep, treacherous dust into which any landing vehicle would sink, with disastrous results. Automatic vehicles had ruled out this theory, but it was still possible that there could be dangerously soft or uneven areas – and if the *Eagle* made a faulty touchdown there could be no hope of rescue. It was a tense time. The astronauts themselves were much too busy to pause and reflect (as they told me later), but the strain in Mission Control, Houston, must have been unbearable. It was bad enough in our London studio.

We followed the space-craft down, and then, at last, came Neil's voice: 'The *Eagle* has landed.' The surge of relief spread from Houston to the whole of the world. I have no clear recollection of what I said; I was on the air 'live', and subsequently the BBC managed to lose all the tapes, so I can only hope that my comments were coherent. The dream of so many centuries had come true.

To me, this was the supreme moment of the whole expedition. Even Neil's later words, as he stepped out on to the lunar surface – 'That's one small step for a man, one giant leap for mankind' – did not have the same impact, even though they will be remembered as long as *Homo sapiens* lasts. The other moment of extreme tension came when the astronauts prepared to blast away from the Moon, to rejoin Collins in lunar orbit. They were entirely dependent upon the single ascent engine of the module; it had to work, properly, first time. Mercifully it did.

Six more Apollos followed that first mission, and all but one were successful; there were no human casualties, and there can be no doubt that the whole programme represented an outstanding triumph – even though it had been initiated not for purely scientific reasons, but by a feverish rush to reach the Moon before the Russians were able to do so. It was to be several decades before space research became truly international.

Way back in 1969 it was widely believed that a fully-fledged Lunar Base would be built within a few years, and that lunar tourism was looming ahead. Now, at the start of the new millennium, these aims are still a long way off. During a *Sky at Night* broadcast in 1970 I was talking to Neil Armstrong about this, and he was definite enough: 'I'm quite certain that we'll have such bases in our lifetime.' Sadly, it is now clear that he was being overoptimistic, and no men have been to the Moon for over 30 years. We must be prepared to bide our time.

When I began studying the Moon, in the 1930s, the whole idea of space-travel was officially dismissed as science fiction, and Interplanetary Societies were put into the same category as the various Flat Earth organizations (flying saucers, of course, lay in the future). Professional astronomers paid scant attention to the lunar surface, and Moon-mapping was left mainly to amateurs, many of whom showed themselves to be extremely competent. The outlook changed during the 1950s, when it became clear that the Moon was not completely out of reach, and today we have a very different outlook. We have surveyed the whole of the Moon, and we have a good understanding of the lunar world.

Moreover, the Moon is less unfriendly than was once believed, despite the lack of atmosphere and the extreme temperature range. There are no choking gases or corrosive acids, as with Venus, or lethal radiation zones, as with Jupiter; natural activity is at a very low level, and the mild moonquakes can do explorers no harm. Above all, there is no trace of life. To run the risk of wiping out any indigenous species would be morally wrong, but this does not apply to the sterile Moon. We are starting to plan potential colonies there, even though there is no chance of turning the Moon into a sort of second Earth.

I am essentially a practical lunar observer, and I make no pretence of being knowledgeable about rocketry; to use scientific slang, I am no 'hardware man'. In this book, then, I will try to present a concise account of the Moon itself, tracing the story of how it has been studied, giving some of the old legends and theories, and leading on to the present day. Much will have to be left out, but there is a great deal to say, and we are learning more every year. For a start, let us look back into the remote past, when the very nature of the Moon was still a mystery.

2

Myth and Legend

Many thousands of years ago, long before the start of recorded history, our ancestors must have been fascinated by the Moon, and wondered just what it was. For part of every month it dominated the night sky; it moved quickly against the starry background; it showed regular if puzzling changes of shape. It took second place only to the Sun. Could it be a god, or at least the home of a god? There seemed no obvious reason why not.

Of course the Moon was useful, both as a source of light during the night and as a timekeeper; it also regulated the tides, as must have become clear at a very early stage. Moon myths probably go back almost as far as the human race, and every country seems to have its own legends. Some of these old tales are charming. For instance, in Old Japan it was believed that a powerful emperor lived on the Moon and looked down on to Earth, making sure that all was well. He even sent his daughter to live on Earth for a while, because he wanted her to understand how important it was to watch over humanity – and today, when you look at the full moon, you can see the princess's face smiling down at you... Certainly it is easy enough to see the bright areas and the dark patches, and there can be nobody who has not heard of the Man in the Moon. According to a North German myth, the Old Man was caught stealing cabbages from his neighbours, and as a punishment was despatched to the Moon, where he can be seen by everybody, 'bearing his load of cabbages to all eternity'. Another version, from the island of Sylt, makes him a sheep-stealer. A third German myth tells how an old man persisted in cutting sticks on a Sunday, and was finally banished to the Moon; he had been given the choice of freezing on the Moon or burning in the Sun and, perhaps wisely, opted for the lunar frost. The Chinese had a different idea; to them the Old Man was to all intents and purposes the god of marriages, and it was his solemn duty to link husband and wife with an invisible silken thread which could not be broken until one of the partners died.

Women, of course, could not be left out. I particularly like a Polynesian legend, according to which a girl named Sina was unwise enough to compare the Moon with a huge bread-fruit – with the result that the Moon reached down and snatched up both Sina and her child, so that they can still be seen in the Moon today. All in all, it was not a good idea to make disparaging remarks about the Moon, as a woman named Rona found out. According to the New Zealand Maori people, Rona was the daughter of the sea-god Tangaroa. One night she was carrying a bucket of stream water back to her children when the Moon slipped behind clouds, and in the darkness Rona stubbed her foot against a root. Instinctively she blamed the Moon – and must have spoken too loudly, because the Moon heard, grabbing both Rona and her bucket before putting a curse on the whole Maori tribe. Woman and bucket are still to be seen; when Rona upsets her bucket, rain falls.

To the people of Greenland, the Sun and Moon were brother and sister, and for some reason or other the Sun rubbed soot in his sister's face. Rather naturally, the Moon chased him – but she can never catch him, because she cannot fly so high. Every few weeks she feels the need for a rest, so she comes back to the ground, climbs into a sleigh drawn by four dogs, and goes seal-hunting. After eating several seals she regains her strength, and the Moon once more becomes full, so that the chase can start all over again.

Iaw is the Moon-god for the Mamaiurans of Brazil; his brother was named Kuat. Originally it was always dark on Earth, because there were so many birds in the sky that sunlight could never percolate through. Tired of this gloom, Iaw and Kuat decided to take action. They hid themselves and managed to capture the bird king, Urubutsin; before releasing him, they had made him promise that he would force the birds to share the daylight with the people of Earth.

To the people of Van, in Turkey, the Moon was a young bachelor, and was engaged to the Sun. Originally the Moon had shone in the daytime and the Sun at night, but the Sun, being a girl, was afraid of the dark, and persuaded her fiancé to change places. In another Turkish tale, the Moon was very fond of his mother and used to follow her about everywhere, greatly to her annoyance. Once he followed her when she was washing dishes, and the mother was so angry that she threw the dishcloth in his face, which at least explains why the Moon's disk looks so stained.

Come now to the Tsimshiam people of the New World. There was once a chief, who had two handsome sons; the younger was known as One Who Walks All Over the Sky, and the elder as One Who Walks About Early. The younger brother was always sad to see the sky so dark, so he made a mask of wood and pitch and lit it. Each day he travels across the sky, casting light on to the Earth, and at night he sleeps below the horizon. When he snores, sparks fly from his mask and make the stars. The older brother became so jealous that he smeared fat and charcoal on his face and made his own way across the sky, becoming the Moon.

Animals found their way to the Moon, not always of their own free will. From India comes the remarkable tale of how a wolf fell violently in love with a toad, and refused to take 'no' for an answer; to escape, the toad jumped on to the Moon and stayed there. (The wolf's comments are not on record.) Also from India we hear how a hare offered to cook himself in order to provide dinner for a hungry Brahmin. The offer was declined with thanks, but the hare was placed in the Moon as a reward for his courtesy.

Yet another Chinese legend explains how there was once a great drought, and a herd of elephants came to drink at a sheet of water called the Moon Lake. They trampled down so many of the local hare population that when they next appeared, a far-sighted hare pointed out that they were annoying the Moon-goddess by disturbing her reflection in the water. The elephants agreed that this was most unwise, and departed hastily.

Moon-worship was widespread in ancient times, and the lunar god was one of the most important of all the deities. Generally – not always – it was male, and only the Sun was more important. The ruins of a large lunar temple have been uncovered at Ur, while the Egyptians had two moon-gods, Khonsu (also the God of Time – here we have a link with the calendar) and Thoth. In Greece, Artemis was the lunar deity, while the Japanese moon-goddess rejoiced in the name of Tsuki-yomo-no-kami. The Aleutian Islanders were in the habit of stoning to death anyone who had been incautious enough to offend the Moon. And from the Confessional of Ecgbert, Archbishop of York, we learn that the British Druids still paid homage to the Moon as late as the eighth century AD.

Quite apart from these myths, the early peoples managed to find out at least something about the Moon itself. At first they believed that it actually changed shape from night to night – in Bushman mythology the Moon was believed to have offended the Sun, and is regularly pierced by the Sun's rays until he pleads for mercy and is gradually restored – but it was then found that this could not be so, because the 'dark' place of the disk can often be seen shining faintly alongside the brilliant crescent. In fact, this effect can be seen almost every night when the Moon is crescent-shaped; some people call it 'the Old Moon in the Young Moon's arms'. As Leonardo da Vinci pointed out later, it is due to light reflected on to the Moon from the Earth. At least it showed that the Moon is always a complete disk.

On the other hand, the ancients had no idea of the plan of the universe, and some of their ideas about the Earth itself were decidedly peculiar. Usually the world was a flat plate, sometimes floating in water and sometimes surrounded by a solid sky. According to the Hindus it stood on the back of four elephants, which in turn rested upon the shell of a vast tortoise swimming in a boundless ocean. One cannot help feeling somewhat sorry for the tortoise, but there were other theories too. One Indian cult taught that the Earth is supported on twelve massive pillars, so that during the hours of darkness the Sun has to thread its way through a kind of maze without touching any of the pillars.

In Egypt the priests, who ranked as the country's leading scholars, made the initial mistake of supposing that the universe takes the form of a rectangular box, with the longer sides running north–south. There is a flat ceiling, supported by four pillars which are connected by a mountain chain; below this chain lies a ledge containing the celestial river Ur-nes, along which sail the boats carrying the Sun, Moon and other gods. When a boat comes to one of the sharp corners of the river, it turns abruptly at right angles and continues blithely upon its way.

All this is intriguing, but it is not science; and true science did not begin until the time of the Greeks. Neither was it quick to develop. The first great Greek philosopher, Thales, was born shortly before 600 BC, while Ptolemy, the last famous scientist of Classical times, died about AD 180. This gives us a total time-span of 800 years, so that chronologically Ptolemy was as far away from Thales as we are from the Crusades.

Thales believed the world to be shaped like a log or cork, and to be floating on water. His younger contemporary, Anaximander, had definite ideas about the Moon, and described it as 'a circle nineteen times as large as the Earth; it is shaped like a chariot-wheel, the rim of which is hollow and full of fire, as that of the Sun also is; it has one vent, like the nozzle of a pair of bellows; its eclipses depend upon the turnings of the wheel.' On the credit side, Anaximander did at least claim that the Earth is suspended freely in space without being held up by pillars, elephants, tortoises or anything else.

It would be too much of a digression to describe many of the other ideas current during the first centuries of Greek greatness, but I cannot resist quoting Xenophanes, who died in or about 478 BC at the advanced age of nearly one hundred. 'There are many suns and moons according to the regions, divisions and zones of the Earth… the Earth is flat. On its upper side it touches the air; on the underside it extends without limit.' Xenophanes believed that the various suns, moons and stars were made of clouds set on fire by some process which he did not describe. At about the same time another philosopher, Heraclitus, wrote that the diameter of the Sun is about twelve inches, which is something of an underestimate!

Gradually, the idea of the Moon as a body moving round the spherical Earth at a relatively slight distance began to gain support. It was also found that it has no light of its own, and depends upon light reflected from the Sun. Around 270 BC Eratosthenes of Cyrene, librarian in charge of the great collection of books at Alexandria, made a remarkably accurate measurement of the size of the Earth, and his contemporary Aristarchus had a very shrewd idea of the Moon's distance from us. (Aristarchus was also one of the first to suggest that the Earth moves round the Sun, but at the time he found few followers.) Lunar eclipses were correctly explained as being due to the Moon's entry into the shadow cast by the Earth, and it was taught that the markings on the disk were due to lofty mountains and deep valleys.

What about the chances of life on the Moon? To some of the Greeks there seemed every reason to suppose that beings of some kind lived there, though whether these beings were human or merely 'spirits' was an open question. Around AD 80 the celebrated writer Plutarch produced *De Facie in Orbe Lunæ* ('On the Face in the Orb of the Moon') in which he maintained that the lunar world is 'earthy', with mountains and ravines; but he was quite convinced that it must be inhabited. Space-travel ideas also go back to the Greeks, though the suggested methods were not of the kind calculated to appeal to NASA or to Neil Armstrong. I will have more to say about them later.

The last great Classical philosopher, Ptolemy, wrote a book in which he summed up the astronomical knowledge of his time. Because his text has come down to us only via its Arabic translation, we usually know it by its Arabic title: the *Almagest*. In it, Ptolemy described how the celestial bodies move round the Earth in a rather complicated manner. Like virtually all his contemporaries he believed that all paths or orbits must be perfectly circular, but he was too good an observer and mathematician to think that the situation was straightforward, and he had to introduce various refinements which need not concern us at the moment. His system – always called the Ptolemaic, though Ptolemy himself did not invent it – was generally accepted for more than a thousand years after his death, but unfortunately it was completely wrong. Astronomy could make real strides only when the idea of a motionless, central Earth had been cleared out of the way.

Actually, the Moon is the only natural body which really does move round the Earth – and even this is an oversimplification. The planets revolve round the Sun, while the stars are suns in their own right. Few of the Greeks had the courage to dethrone the Earth from its proud central position (Aristarchus was an exception), and the great upheaval in human thought was delayed until less than five hundred years ago, which is a sobering thought. I am trying to do no more than give a brief sketch of events, so we can safely skip more than a dozen centuries and come to what is generally called the Copernican Revolution.

It began in 1543, with the appearance of a book, *De Revolutionibus Orbium Cælestium* ('Concerning the Revolutions of the Celestial Orbs'). The author – Copernicus, a Polish canon – prudently held back publication until he was dying, because he was well aware that the Christian Church would not take kindly to the idea of a moving Earth; and in this he was correct. The 'Copernicans' were savagely persecuted, and one of them, Giordano Bruno, was burned at the stake in Rome in 1600. What really tipped the scales was the work of a German mathematician, Johannes Kepler, who spent many years in studying observations of the planet Mars made by Tycho Brahe, an eccentric Dane who was firmly committed to the idea of a central Earth but whose measurements of the positions of the stars and planets were amazingly accurate. Kepler found that the planets move round the Sun not in circles, but in ellipses. His famous Laws of Planetary Motion, published between 1609 and 1618, gave the death-blow to Ptolemy's theory. It was during this period, too, that telescopes were turned towards the sky; for the first time men could see that instead of being covered with grassy plains, glittering oceans and spreading forests, the Moon was a world of rugged mountains, broad plains and huge craters.

Apparently the first man to look seriously at the Moon through a telescope was Thomas Harriott, one-time tutor to Sir Walter Raleigh, but the real pioneer was Galileo, who described the lunar surface in great detail and even made some reasonably successful attempts to measure the heights of the mountains. Galileo's work may be said to have ushered in the 'modern' era, though the old ideas were far from dead. Despite the detailed maps which were drawn up during the following years we still find Sir William Herschel, one of the greatest of all observers, maintaining that the Moon was certainly inhabited – and Herschel died as recently as 1822. But as time went by, and the airless, hostile nature of the Moon became painfully obvious, opinions changed. Well before the first rockets were sent there, the idea of lunar creatures had been firmly relegated to the scientific scrap-heap.

One thing had always been obvious: the Moon is the closest body in the sky, and would have to be our first target. Ideas of lunar travel go back for almost two thousand years, but only in our own time has space-flight become possible. The Apollo astronauts showed the way, and by now we can start serious planning for a Lunar Base. But before we go any further, it seems only right to pause momentarily, and try to put the Moon in its proper perspective with regard to the Earth and the other worlds around us.

3

The Moon in the Solar System

The Moon is so splendid in our skies that we naturally tend to think of it as important. Obviously it cannot rival the Sun, and the difference is greater than most people realize. Sunlight is over half a million times more powerful than the radiance of the full moon, and it is quite wrong to suggest that a moonlit night can be almost as bright as day. Yet the Moon is both beautiful and imposing, and it means much more to us than the tiny, twinkling stars.

Modern astronomy paints a very clear picture. Things are emphatically not what they seem, and of all the celestial bodies visible with the naked eye the Moon is the least important (unless we count cosmical débris and our own artificial satellites). Officially the Moon is not ranked as a planet, and is relegated to the status of a mere junior attendant of the Earth. I am not at all sure that this view is justified, for reasons I propose to give below, but it is true that even Mercury, the smallest of the main planets, is larger and more massive than the Moon.

What makes the Moon unique is its closeness to us, and the fact that it stays with us all the time as we travel round the Sun. Its average distance from the Earth is less than a quarter of a million miles*, and the nearest planet – Venus, incidentally, not Mars – is always a hundred times as far away as this. The distance of the Sun is 400 times that of the Moon. And when we come to consider the stars, we are faced with distances so great that nobody can really appreciate them.

Can you really imagine what is meant by 'a million miles'? I admit that I cannot, and a million miles is a very short stretch on the scale of the universe. Astronomers cannot appreciate great spans of space and vast stretches of time any better than laymen – the only difference is they do not make the mistake of trying! We know that our figures are accurate, and we simply have to accept them.

The Solar System is made up of one star (the Sun), the nine known planets, and various lesser bodies, including the satellites of some of the planets. Only the Sun has light of its own, so that the other members of the system merely

---

* Miles or kilometres? In this book I have decided to use Imperial measures rather than Metric, so that everybody can understand them.

reflect the solar rays in the manner of large but, in general, not very efficient mirrors. (The Moon is particularly poor as a reflector of light: in scientific parlance, it has an average albedo of only 7 per cent.) As a start, it may be helpful to summarize the main details of the planets, which is best done by means of a table:

PLANET	MEAN DISTANCE FROM THE SUN, MILLIONS OF MILES	REVOLUTION PERIOD	AXIAL ROTATION PERIOD	DIAMETER, MILES, EQUATORIAL	NUMBER OF SATELLITES
Mercury	36	88 days	58.6 days	3,030	0
Venus	67	225 days	243 days	7,520	0
Earth	93	365 days	23h 56m	7,926	1
Mars	141.5	687 days	24h 37m	4,220	2
Jupiter	483	11.9 years	9h 50m	89,400	17
Saturn	886	29.5 years	10h 14m	74,900	18
Uranus	1,783	84 years	17h 14m	31,800	20
Neptune	2,793	164.8 years	16h 7m	31,400	8
Pluto	3,666	248 years	6 days 9h	1,444	1

The Solar System is divided into two distinct parts. First come four relatively small, solid planets: Mercury to Mars. Next there is a wide gap, filled by thousands of dwarf worlds known as minor planets or asteroids; beyond come the large planets, two 'gas giants' (Jupiter and Saturn) and two 'ice giants' (Uranus and Neptune). Pluto is very much of a maverick. It does not fit into the general scene, and there are serious doubts as to whether it really deserves to rank as a proper planet.

The Moon's diameter is 2,160 miles. Its surface is mountainous, and scarred with craters – which I mention at the outset because they are so significant. It takes 27.3 days to complete one journey round the Earth, and it spins on its axis in exactly the same time – a point to which I return later.

In our lightning tour of the Solar System we must begin with Mercury, named by the ancients after the fleet-footed messenger of the gods. It is never prominent, because it stays in the same part of the sky as the Sun, and is visible with the naked eye only at its best – either low in the west after sunset or low in the east before dawn. One space-craft has flown past it, Mariner 10 in 1974–5, and sent back close-range pictures showing a cratered surface not unlike that of the Moon. Like the Moon, it has almost no atmosphere, and certainly there can be no life here.

Venus, next in order of distance from the Sun, is very different; it can be bright enough to be seen with the naked eye in broad daylight, and can cast shadows. It is about the same size as the Earth, but is a very curious world indeed. It has a dense atmosphere made up chiefly of the heavy, unbreathable gas carbon dioxide, and its clouds are rich in sulphuric acid; the surface temperature is not far short of 1,000 degrees F, and the atmospheric pressure on the surface is about 90 times that of the Earth's air at sea-level. Space-craft studies have shown uplands, craters, high plateaux and volcanoes which are almost certainly active. Telescopically it shows phases, or apparent changes of shape from new to full, like those of the Moon, but ordinary telescopes cannot see through the clouds. It spins slowly on its axis, and technically speaking its 'day' is longer than its 'year'; moreover, Venus turns from east to west, not west to east in the same sense as the Earth. If you could go there, the Sun would rise in the west − but in fact it would never be seen at all, because of the cloud cover; there is no such thing as a sunny day on Venus. Life there is surely impossible. Any astronaut unwise enough to land on the surface would promptly be squashed, corroded, poisoned and fried. Venus was named after the Goddess of Beauty, but conditions there are much more akin to the conventional idea of hell.

Planet number three, our Earth, is the only world in the Solar System which is suited to the existence of intelligent life (whether intelligent life has actually appeared here must be regarded as questionable). Next we come to Mars, whose red colour led to its being named in honour of the God of War. It can be brilliant − brighter than any other celestial body apart from the Sun, the Moon and Venus − and its ruddy hue makes it identifiable at once. It is smaller than the Earth, with a much thinner atmosphere, and it is cold. The poles are covered with white ice-caps which wax and wane with the Martian seasons; much of the surface is covered with red deserts, made up of 'rusty' minerals rather than sand; and there are permanent dark areas, where the red material has been blown away by winds.

Mars is less unlike the Earth than any other planet, and less than a century ago it was widely believed that advanced life-forms might exist there. Some astronomers even drew features which they believed to be artificial canals, built by local inhabitants to make up a vast irrigation system. Alas, the canals do not exist − they were due to tricks of the eye − and there are no Martians. Neither are the dark regions covered with vegetation, and as yet we have no evidence of any life there, though very primitive organisms may exist.

Many unmanned space-craft have been sent to Mars, and we now have detailed maps of the entire surface. Craters are plentiful, and there are massive volcanoes, probably the largest anywhere in the Solar System; one, known as Olympus Mons, is three times as lofty as our Everest, and is crowned by a 40-mile caldera. Active vulcanism belongs to the past; there must once have been surface water, but Mars today is an arid planet. Dust-storms are common, and there are high, ice-crystal clouds, though the atmospheric pressure is nowhere as high as 12 millibars. There are two midget satellites, Phobos and Deimos, neither of which is as much as 20 miles in diameter. Almost certainly they are captured asteroids rather than bona-fide satellites; they are quite unlike our relatively massive Moon.*

In size and mass, Mars is intermediate between the Earth and the Moon. Go there, and you will have only one-third of your Earth weight, because the gravitational pull of Mars is much weaker than ours. This is why the Martian atmosphere is so thin; most of any 'air' originally present has leaked away into space.

One important fact is that all these inner worlds – Mercury, Venus, Mars and also the Martian satellites – are cratered. It is the Earth which is exceptional, inasmuch as our own craters are relatively few and small. There are excellent reasons for this, but for the moment let us press on to look at those regions of the Solar System beyond the orbit of Mars.

The asteroids, or minor planets, are tiny worlds; only one (Ceres) is as much as 500 miles in diameter, and only one (Vesta) is ever visible with the naked eye. More than ten thousand have had their orbits worked out, and the total membership of the swarm may be as high as 100,000. When the first space-craft to Jupiter were sent out, many people had the uneasy feeling that passing through the asteroid belt would be a distinctly hazardous business, but many probes have since made the crossing safely, so that the danger may be less than was originally feared. All the asteroids combined would not make up one body nearly as massive as the Moon.

Several asteroids have been imaged by space-craft; they are – predictably – rocky and cratered; many are irregular in shape, and there must be frequent collisions between minor members of the swarm. One or two of them are known to have satellites of their own; thus Ida, with a longest diameter of less than 40 miles, has a mile-wide attendant, now named Dactyl.

The main belt of asteroids lies between the paths of Mars and Jupiter, but there are some which do not conform. The 'Trojan' asteroids move in Jupiter's orbit, but keep prudently well ahead of or well behind the Giant Planet, and are in no danger of being swallowed up. There are also asteroids which swing closer in than the main swarm, and may pass close to the Earth. In 1994 one

---

* For a full account of the Red Planet, see *Patrick Moore on Mars*, uniform with this book. I hasten to add that I have not actually been to Mars; I have merely looked at it!

midget, as yet unnamed, sped by at a mere 60,000 miles. These NEAs (Near Earth Asteroids) are probably very numerous; most of them are very small indeed – only a few tens or hundreds of yards across – but an asteroid impact would cause widespread devastation, and there are serious suggestions that such an event, around 65 million years ago, led to a change in climate which proved fatal to the dinosaurs. The danger of a major strike in the foreseeable future is slight, but it is not nil. I will have more to say about this when we come to consider the craters of the Moon.

Next in order come the four giants of the Sun's family. Jupiter and Saturn have rocky cores, overlaid by layers of liquid hydrogen and then by the gaseous atmospheres which we can see; the atmospheres are made up mainly of the two lightest gases, hydrogen and helium. Jupiter's surface features are changing all the time, but there are always dark cloud belts and bright zones; there is also the Great Red Spot, now known to be a whirling storm, whose surface area is greater than that of the Earth. The quick rotation – the Jovian 'day', less than 10 hours long – means that the globe is obviously flattened. Space-craft rendezvous have shown that Jupiter is associated with dangerous radiation zones which would be promptly fatal to any astronauts foolish enough to venture within 100,000 miles of the cloud-tops. Jupiter is a world to be viewed from a respectful distance.

Saturn, smaller than Jupiter and much further from the Sun, is of the same general type, but is distinguished by its magnificent system of rings. The rings are made up of millions of tiny icy particles, whirling round Saturn in the manner of dwarf satellites; they measure almost 170,000 miles from one side to the other, but are less than a mile thick. When best placed they are easy to see with a small telescope, and it is probably fair to say that Saturn is then the most beautiful object in the entire sky. In fact all the giant planets have ring systems, but only Saturn's is bright.

Both Jupiter and Saturn have satellite families. Of Jupiter's attendants, four are large; they are named Io, Europa, Ganymede and Callisto, and are collectively referred to as the Galileans, because their movements were first studied by the great Italian scientist Galileo as long ago as 1610. Europa is slightly smaller than our Moon, Io slightly larger, and Ganymede and Callisto much larger – indeed Ganymede is larger than the planet Mercury, though less massive. All four have points of special interest. Ganymede and Callisto are icy and cratered, Europa icy and smooth, and Io violently volcanic, with eruptions going on all the time. The remaining thirteen satellites are very small, and may well be captured asteroids.

Saturn has only one large satellite, Titan, which is unique among satellites inasmuch as it has a dense atmosphere, made up of a mixture of nitrogen and methane. A space-craft, Cassini, is at present on its way there, and if all goes well will drop a probe on to the surface in the year 2004. The nature of the surface of Titan is not known; there may be chemical lakes or even oceans. Certainly Titan is unlike any other body in the Solar System. All the other members of Saturn's family are below 1,000 miles in diameter.

The two outer giant planets, Uranus and Neptune, are made up mainly of 'ices', including water; their atmospheres consist largely of hydrogen. Uranus, discovered in 1781, is just visible with the naked eye if you know where to look for it, while Neptune, first identified in 1846, can be seen with binoculars. Both were surveyed by the Voyager 2 space-craft, Uranus in 1986 and Neptune in 1989; Uranus is rather bland, but Neptune is decidedly more dynamic, with belts, spots and violent winds. The satellites of Uranus are rather undistinguished, but Neptune's senior attendant, Triton, is remarkable. It is smaller than the Moon, but has an extensive though very thin atmosphere; the poles are covered with snow – not water-ice snow, but nitrogen snow; the temperature is so low that nitrogen freezes out on to the surface. There are also active nitrogen geysers.

The ninth planet, Pluto, was discovered as recently as 1930. It is too faint to be seen with a small telescope, and it seems to be in a class of its own. It has a very eccentric orbit; it takes 248 years to make one circuit of the Sun, and for part of this time it is actually closer in than Neptune, though its path is appreciably tilted, and there is no fear of collision. Pluto is smaller than the Moon, and has a companion, Charon, whose diameter is more than half that of Pluto itself. Another peculiarity is that Charon's orbital period, 6 days 9 hours, is the same as Pluto's axial rotation period; to an observer on Pluto, Charon would remain in a fixed position in the sky. Both would be gloomy places; even the Sun would appear as nothing more than a starlike though very brilliant point.

There may be another planet, much further from the Sun, but we have no definite evidence of its existence. There are, however, many much smaller bodies in these remote regions, making up what might be regarded as an outer asteroid zone; it is known as the Kuiper Belt in honour of the Dutch astronomer Gerard Kuiper, who first suggested that it might be present. It is quite possible that Pluto and Charon are merely the largest members of the Kuiper Belt.

Comets are the most erratic members of the Solar System. They are not solid and rocky, and are not nearly so important as they sometimes look. They move round the Sun, but most of them do so in very elliptical paths; their periods range from just over 3 years to many centuries. They are flimsy things, and are made up largely of ice; indeed, it has been said that a comet is best described as a 'dirty ice-ball'. The nucleus is seldom more than a few miles in diameter. Because all comets lie far beyond the Earth's atmosphere, and are millions of miles away, they cannot be seen to shift perceptibly against the starry background; if you see something crawling obviously across the sky, it cannot be a comet.

When a comet nears the Sun, its ices begin to evaporate, and the comet may develop a long tail or tails. Now and then we have really spectacular visitors; for example the comets of 1811, 1843 and 1882 cast strong shadows. Really bright comets were rare during the twentieth century, but in 1997 we did have Comet Hale–Bopp (named after its two discoverers, Alan Hale and Thomas Bopp), which was prominent for many weeks. If you missed it, I fear that you will have a long wait before you can see it again, because it will not be back to the neighbourhood of the Sun for about 2,400 years. There are plenty of short-period comets within periods of a few years, so that we always know when and where to expect them, but few attain naked-eye visibility, and not many of them have tails. The only bright comet with a period of less than a century is Halley's, which was last at perihelion (that is to say, its closest point to the Sun) in 1986, and will be back on schedule in 2061.

There used to be a general fear of comets, mainly because of the consequences of a direct collision, which would certainly do a great deal of damage. In 1994 a comet was indeed seen to hit Jupiter, and produced effects in the Jovian atmosphere which persisted for months. Had that particular comet struck Earth, the results would have been devastating indeed.

Meteors are cosmical débris – tiny particles, usually smaller than pins' heads, left behind by comets. If one of these particles dashes into the upper air, travelling at up to 45 miles per second, it becomes heated by friction, and destroys itself in the streak of luminosity which we call a shooting star, ending its journey in the form of fine 'dust'. Meteors tend to travel round the Sun in shoals, and when the Earth passes through a shoal we see a shower of shooting stars. This happens many times each year. The August meteors, known as the Perseids, can always be relied upon to produce a good display; go out on a clear, dark night at any time during the first fortnight in August, and you will be unlucky not to see several Perseids. There are also non-shower or sporadic meteors, which may appear from any direction at any moment.

Meteorites are quite different, and are not associated with comets. They are too massive to be burned away as they plunge earthward, and sometimes produce craters. Most museums have collections of meteorites which range in size from tiny pebbles to large blocks, and occasional giants have been found. The largest meteorite 'in captivity', so to speak, weighs 36 tons, and you can see it if you go to the Hayden Planetarium in New York; it was discovered in

Greenland by the polar explorer Robert Peary. The best-known meteorite to have fallen in Britain in recent years landed in the Leicester village of Barwell on Christmas Eve 1965. It broke up during descent, and fragments of it were scattered over a wide area. Before its disruption, the meteorite might have had a diameter of at least six feet.

Major falls are rare, and we know of only two in the last few hundred years. The first was that of 1908, in the Tunguska region of Siberia, which blew pine-trees flat for miles around, though mercifully the region was uninhabited and there were no casualties (except among the local reindeer population). No meteoritic fragments have been found, and it seems that the missile was icy in nature, perhaps a piece of a small comet. This would account for the lack of débris, because the icy materials making up a comet would have evaporated quickly. There is no doubt about the second fall, in 1947 not far from Vladivostok; many meteoritic fragments were collected, and there were numerous small craters.

Other proven meteoritic falls of this magnitude are prehistoric. The holder of the heavyweight record is the Hoba West Meteorite, which is still lying where it fell near Grootfontein in Southern Africa; nobody is likely to run away with it, since it weighs at least 60 tons. You may also be interested to know that the Sacred Stone at Mecca is certainly a meteorite, though we have no idea when it fell.

Go to Arizona, not too far from the town of Winslow, and you will find the famous Meteor Crater – really it should be Meteor*ite* Crater – which is almost a mile across, and of whose nature there is no doubt at all. It was formed around 50,000 years ago, well before any people lived there. The Wolf Creek Crater in Australia is also of impact origin, and by now we have identified many others. I will have much more to say about impact craters when we discuss the surface of the Moon, which is crowded with them.

There is a fundamental difference between meteorites and shooting-star meteors. On the other hand it now seems that there is no real distinction between meteoritic bodies and asteroids – and there is little doubt that meteorites in general come from the asteroid zone, though there are suggestions that some of them may have been blasted away from the Moon or even Mars.

Even apart from meteoritic particles, the space between the planets is not empty. There is a great deal of thinly-spread material, lying chiefly in the main plane of the Solar System and causing the faint glows which we call the Zodiacal Light and the Gegenschein.

The Sun is, of course, the ruler of the Solar System, and it is of supreme importance to us, even though we now know that it is nothing more than a very ordinary star. It is large – you could cram over a million bodies the volume of the Earth inside it, and still leave room to spare – and it is hot; even the surface is at a temperature of well over 5,000 degrees C, and near the centre of the globe, where the energy is being produced, the temperature soars to the incredible value of about 15 million degrees. Hydrogen is the Sun's 'fuel'; near the core, hydrogen is being converted into another gas, helium, with release of energy and loss of mass. It is this energy which makes the Sun shine, and the loss in mass amounts to four million tons per second. This may seem a great deal, but the Sun is at least 5,000 million years old, and it will shine as much as it does now for several thousands of millions of years yet, though it will not last for ever.

The stars are so remote that ordinary units of measurement, such as the mile and the kilometre, are inconveniently short. Instead, we usually measure these vast distances in light-years. Light travels at a speed of 186,000 miles per second, so that in a year it can cover almost 6 million million miles – and this is termed a light-year, which, please note, is a measure of distance, not of time. Light can leap from the Moon to the Earth in $1\frac{1}{4}$ seconds, and from the Sun to the Earth in 8.6 minutes, but the light from the nearest star beyond the Sun takes over 4 years to reach us. This is equivalent to around 24 million million miles. We see that star as it used to be over 4 years ago; look at the Pole Star, and you see it as it was 680 years ago, when Edward II was King of England. Once we look beyond the Solar System, our view of the universe is bound to be very out of date.

All the stars are suns, and many of them are far larger, hotter and more luminous than ours; for example, the brilliant white Rigel, in Orion, could match at least 60,000 Suns, but it is 900 light-years away. On the other hand there are also many stars which are much less powerful than the Sun. If we represent the Sun by an ordinary electric light bulb, the most luminous stars will be searchlights, while the feeblest will be dim glow-worms.

All the stars are racing about at high speeds, but they are so far away that their individual patterns or 'constellations' do not change markedly over many lifetimes; go back to the time of King Canute, Julius Cæsar or the builders of the Pyramids, and the constellations would look virtually the same as they do now. Only the members of the Solar System are close enough to show obvious shifts in position over short periods.

Our own system of stars, the Galaxy, contains around 100,000 million stars. It is a flattened system – shaped rather like two fried eggs clapped together back to back – and when we look along the main plane of the system we see many stars in almost the same lines of sight, producing the effect of the Milky Way. There is no longer any doubt that many stars have planet-families of their own, so that our Solar System is by no means unusual. Even this is no more than a start; millions of light-years away we can see other galaxies, just as populous as ours. The furthest galaxies so far recorded are at least 12,000 million light-years from us.

Life? Well, there is every reason to suppose that life is widespread in the universe, even though we have as yet no positive proof. Yet the nearest extra-terrestrial civilizations must be light-years away, and the only possible way to contact them, at our present stage of development, is by radio, since radio signals travel at the same speed as light. The nearest solar-type stars are about 11 light-years away, and sending out rockets to them is out of the question. If we are to achieve interstellar travel, it must be by some method about which our present ignorance is complete. I do not claim that it will never be done; I do say that we need some dramatic 'breakthrough', which may not be for a long time yet.

I do not apologize for this somewhat lengthy digression, because it serves to show that we must avoid being parochial. The Solar System is unimportant in the universe as a whole, and the Moon is a junior member of it. On the other hand it is of special interest to us, and the more we learn about it the more fascinating it becomes.

4

The Origin of the Moon

How did the Moon come into being? Did it break away from the Earth, or was it once an independent body? We have to admit that we are not certain, but at least we now believe that we may be on the right track.

The whole question is linked with that of the origin of the Solar System itself. We have to start somewhere, and one thing we do know, with fair certainty, is the age of the Earth, which proves to be about 4,600 million years. There are various lines of investigation, all of which lead to much the same result. One of the most convincing methods of attack concerns the decay of certain heavy elements, such as uranium. Uranium is radioactive, and decays gradually into lead, but it is in no hurry; its half-life – that is to say, the time taken for half the original uranium to change into lead – is over 4,000 million years. Fortunately, the lead produced by uranium decay is not quite the same as ordinary lead, and Nature has presented us with a very powerful research tool, because the quantity of uranium-lead associated with the remaining uranium tells us how long the process has been going on. (I realize that this is a gross oversimplification, but it will suffice for the moment.) The oldest Earth rocks date back well over 4,000 million years, and so the world itself must be older than this.

In 1796 the French astronomer Pierre Simon de Laplace proposed what was later called the Nebular Hypothesis, according to which the Solar System began as a vast, slowly-rotating gas-cloud. As it shrank under the influence of gravitation, various rings broke off, each of which condensed into a planet; the outer planets were therefore produced first, while the inner planets (Mars, Venus, Earth and Mercury) were younger. The central part of the original cloud became the Sun; the Moon was born from a gaseous ring thrown off by the contracting Earth.

The Nebular Hypothesis seemed very convincing, but mathematicians had their doubts about it, and eventually they showed, beyond all reasonable doubt, that Laplace's theory in its original form simply would not work.

Next, from 1901, came the 'tidal' theories, according to which the planets were pulled off the Sun by the action of a passing star. A long, cigar-shaped tongue of material would be torn away, and the planets condensed out of this tongue – the largest planets, Jupiter and Saturn, from the thick middle part of the cigar. Again there were found to be fatal mathematical objections, and various other theories produced during the first part of the twentieth century were no better.

Modern ideas are much more akin to Laplace's than to tidal hypotheses. It is thought that the Solar System began as a huge dust-and-gas cloud, part of which started to collapse and rotate, much as Laplace had supposed, but there were no rings; a 'solar nebula' was formed, and in a relatively short time, perhaps as little as 100,000 years, turned into a fledgling star. The temperature rose, and the solar nebula was forced into the form of a flattened, rotating disk. The central temperature rose still further, and the proto-Sun became a true star; it sent out a sort of 'stellar wind' which forced outward the lightest gases, hydrogen and helium. The planets built up from the solar nebula by the process of accretion. The inner, rocky planets lost their hydrogen and helium, but further out, where the temperature was much lower, Jupiter and Saturn grew quickly enough to pull in material from the surrounding nebulosity; Uranus and Neptune, slower to form, could not do the same, because by the time they had become sufficiently massive the solar nebula had more or less dispersed. This is why Uranus and Neptune contain less hydrogen and helium than Jupiter or Saturn, but more 'ices'.

In the early stages there was a great deal of material 'left over', so to speak, and the planets were subjected to heavy bombardment, which explains the cratering – particularly evident in the case of the Moon. The main bombardment ended around 4,000 million years ago, and eventually the Solar System assumed its present form.

Armed with this information, we can start to look at the various theories about the birth of the Moon. Basically, there are four: fission, condensation, capture and giant impact.

The fission hypothesis was originally proposed in 1881 by George Darwin, son of the great naturalist Charles Darwin. It has long since been discarded, but it held sway for many years. Darwin started by assuming that the Earth and the Moon originally made up one body, and that the Moon was thrown off as a fluid mass. In a modified version of this idea, the Earth had cooled down sufficiently to form a thin crust before the separation took place, and the sequence of events was worked out in great detail. The Earth rotated rapidly on its axis, as in the state known as 'unstable equilibrium', so that it became egg-shaped, spinning about its shorter axis. Two main forces were acting upon it: the tides raised by the Sun, and its own natural period of vibration. When these two forces were in resonance (that is to say, acting together) the tides increased to such an extent that the whole body became first pear-shaped and then dumbbell-shaped, with one 'bell' (the Earth) much larger than the other (the future Moon). Eventually the neck of the dumbbell broke, and the Moon moved away, settling into a stable orbit.

All this seemed rational enough, and was supported by many leading astronomers; one of these was an American, W. H. Pickering, who was particularly interested in the Moon and who produced one of the first good lunar photographic atlases. Following up an earlier suggestion by a geologist, Osmond Fisher, Pickering went even further than Darwin had done, and wrote that if the theory were correct the thin crust of the otherwise fluid Earth must have been torn apart, leaving a huge hollow where the thrown-off mass had once been. Moreover, the shock caused by the final fracture would have been violent enough to crack the fragile crust in other places as well.

This fitted in with the theory of continental drift, originally proposed by a meteorologist named Alfred Wegener – severely criticised for some time, but now universally accepted. A glance at any map of the Earth's globe shows that if the opposite sides of the Atlantic Ocean were clapped together they would fit into each other. Allowing for the sea having washed away portions of land here and there, and supposing Britain and Europe to be joined – as they were, only 10,000 years ago – the relationship is striking. The bulge of South America fits into the hollow of Africa, and the eastern coast of North America with the west coast of Europe. The continents may be regarded as 'rafts', floating around very slowly above the mantle below. Pickering looked at the Pacific Ocean, which is very roughly circular. He believed that the rounded hollow which now forms the bed of the Pacific was nothing more nor less than the scar left by the departing Moon, so that our satellite was born in the spot where our greatest ocean now rolls.

It is easy to continue the story. Pickering went on to explain that the crust of the Earth cracked under the shock, and portions of it floated apart, to settle eventually in the places where we now find Eurasia and the Americas. The crustal cracking exposed the intensely hot interior of the Earth, and the fragments of crust floated as skin or scum on the molten globe. The lava surface exposed beneath the broken pieces of crust cooled and solidified, and later, when water was able to condense on the surface, became the Atlantic Ocean. What could be neater?

Alas, it has become clear that there are fatal weaknesses in the whole theory. First, the Pacific is only a few miles deep, and cannot represent the scar left by the hurling of a body the size of the Moon. Represent the Earth by a tennis ball, and the depth of the Pacific will be less than the thickness of a postage stamp – whereas the Moon would be around the size of a table-tennis ball. Neither does it seem likely that a large mass could be thrown off in the way that Darwin and Pickering supposed. All in all, we have to agree that the fission theory fails.

Come next to the condensation theory, supported by one of the last century's great geologists, Harold Urey. According to this scenario, the Earth and the Moon were formed from the solar nebula at the same time and in the same region, so that they have always remained gravitationally linked. Certainly this ties in well with my contention that the Moon should be classed as a companion planet rather than as a satellite. Admittedly there are four planetary satellites which are larger than the Moon – three in Jupiter's family, and one in Saturn's – but these move round giant planets, and are very small and lightweight compared with their primaries. The Moon has $\frac{1}{81}$ the mass of the Earth. The mass ratio between Jupiter and its largest satellite, Ganymede, is 17,700 to 1; between Saturn and Titan, it is 4,150 to 1. Apart from Earth, there is no small planet with a large satellite (I exclude the Pluto/Charon pair, which is in a different category altogether). It may or may not be significant that the ratio between the Earth and the Moon is not wildly different from that between Venus and Mercury.

Yet there are very serious weaknesses in the condensation theory. In particular, the compositions of the Earth and the Moon are not the same. The average density of the Moon's globe is much lower, and it contains much less iron. There are ways to avoid this difficulty, and there is still wide support for the condensation picture; at least it is a serious contender.

Suppose that the Moon were formed in a different part of the solar nebula, and was captured by the Earth much later? This seems very unlikely, for various reasons; such a capture would involve a set of special circumstances. The theory cannot be entirely ruled out, but very few people nowadays have any faith in it. So we turn finally to a completely different idea, proposed in 1974 by the American astronomers W. Hartmann and D. R. Davis. This involves a giant impact.

Hartmann and Davis suggest (as Darwin did) that the Earth and the Moon originally formed one body. About 4,500 million years ago this proto-Earth was struck by a massive impactor, perhaps as large as Mars. The crust and mantle of the proto-Earth were blown off; part of the débris fell gradually back to Earth, but the rest condensed into the present Moon. Since the Moon was formed from the lower-density crust and mantle of the original body, this does at least explain its lower density. It also explains why the Earth and the Moon are of the same age. So far no mathematician has been able to come up with any fatal objections, and it is probably true to say that the giant impact theory is the most popular at the moment, even though we have to admit that we do not know nearly as much about the Moon's origin as we would like to do. (Harold Urey once made the caustic comment that because all theories of lunar origin seemed unlikely, science had proved that the Moon cannot exist!)

There seems little doubt that after its formation the Moon was much closer to us than it is today. If so, then the 'month', the time taken for the Moon to complete one orbit, must have been shorter; at a distance of 12,000 miles the orbital period would have been no more than $6\frac{1}{2}$ hours. In this kind of situation, the tides raised by the two bodies upon each other would have been violent indeed, and the Moon, so much less massive than the Earth, was the more affected. These mutual tides had two important results. They slowed down the axial rotation periods of both Earth and Moon, and they pushed the Moon further away.

As the torn, tide-rent Moon receded, the persistent pull of the Earth raised a permanent 'bulge' in the lunar globe, which had not yet solidified; the Earth tended to keep this bulge turned toward it, so that the Moon's rotation was slowed down. The effect may be compared with that of a cycle wheel rotating between two brake shoes, though I admit that this analogy should not be taken too far. The Moon had to fight against the tidal forces which were braking its rotation, and the process went on until, relative to Earth, the Moon's rotation had stopped altogether. By then the distance had increased to its present value; the revolution period was 27.3 days, and the time taken for the Moon to spin on its axis was exactly the same. Meanwhile, the Earth's axial rotation period had increased to almost 24 hours.

Because the Moon's orbital period is equal to its rotation period, the Moon keeps the same face turned towards us all the time – a point to which I will return later; this state of affairs is known as captured or synchronous rotation. Note, however, that the Moon does not keep the same face turned sunward, as has sometimes been supposed. Day and night conditions are the same everywhere over the lunar surface, though from the far side, of course, the Earth can never be seen.

Perhaps I had better pause to say a little about why these tidal effects drive the Moon away. The crux of the matter is what is termed angular momentum. The angular momentum of a moving body is obtained by multiplying together its mass, the square of its distance from the centre of motion, and the rate of angular motion – that is to say the rate of axial rotation. According to a well-known principle, angular momentum can never be destroyed; it can only be converted. If therefore the axial rotation is slowed down, as happened in the Earth–Moon system because of tidal forces, something else had to increase, and this 'something' was the distance between the two bodies.

The process is not complete even now, because the Moon's tidal pull on the Earth is still braking our rotation. Each day is approximately 0.00000002 second longer than its predecessor, though there are also irregular fluctuations unconnected with the Moon. Also, the Moon is still receding from us. The rate of increase is only 1½ inches per year, however, so that we need be in no hurry to study the Moon before it disappears into the distance!

In fact, the Moon will not go on receding indefinitely. If it could move out to 350,000 miles it would start to draw inward again, because of tidal effects due to the Sun, and would eventually be broken up into a swarm of particles – but this will not happen, because long before the critical period both Earth and Moon will have been destroyed as the Sun swells out to become a red giant star. Meanwhile, we may be sure that nothing dramatic will happen to the Moon yet awhile.

Of the other inner planets, Mercury and Venus are solitary travellers in space. The two dwarf moons of Mars are almost certainly captured asteroids. Is it therefore possible that the Earth has any small attendants of this kind?

There is certainly no junior satellite within reasonable range. At a distance equal to that of the Moon, a 25-mile body of average reflecting power would shine as a brilliant star, and a satellite only one mile across would be visible in binoculars even if it were made up of darkish rock. At two million miles a 25-mile body would be seen with the naked eye, and a one-mile satellite would be detectable with equipment used by the average amateur. Photographic surveys of the sky are now so complete that anything of this sort would have been tracked down years ago. If a minor satellite exists, it must be either much further away or else very small indeed.

In the nineteenth century, a French writer named Petit published a paper about a suggested second satellite moving at a distance of 1,650 miles from the Earth's surface, and with a period of 3 hours 20 minutes. I am not sure how Petit arrived at this conclusion, which is certainly wrong, but the idea was taken up by Jules Verne in his famous novel *Round the Moon*, published in 1870. I will say more about Verne later; meanwhile, it is worth repeating his description of what was seen by one of his fictional space-travellers, President Barbicane, observing through the porthole of the equally fictional moon-projectile:

> *As Barbicane was about to leave the window… his attention was attracted by the approach of a brilliant object. It was an enormous disk, whose dimensions could not be estimated. Its face, which was turned earthward, was brightly illuminated; it might have been taken for a small moon reflecting the light of the large one. It advanced very rapidly, and seemed to be following an orbit round the earth which would intersect the path of the projectile. It moved along in its orbit, and at the same time spun on its own axis… The object grew enormously, and the projectile seemed to be rushing into its path… The*

travellers instinctively recoiled. Their alarm was great, but it did not last long. The object passed within a few hundred yards and vanished, merging into the absolute blackness of space.

It is a wonderful description, and the second satellite was essential in the plot of Verne's story, but it was nothing more than that. Subsequently Clyde Tombaugh, the American astronomer who discovered the planet Pluto in 1930, made a really systematic search for minor Earth satellites, and found none. He used the 24-inch refractor at the Lowell Observatory in Arizona; this was ideal for the search, and Tombaugh's experience was unrivalled. He told me that he was sure there could be no orbiting body worthy to be ranked as a satellite. Of course, there may well be tiny meteoric bodies moving round us at distances greater than that of the Moon, but they must be so small that they must be classed as cosmical débris.

There is also an intriguing suggestion made by the Polish astronomer K. Kordylewski. It goes back to 1961, and remains unconfirmed, but it is worth noting.

Kordylewski began his research for minor satellites in 1951, using the equipment at Kasprowy Wierch and Lomnica in the mountains of Poland. On 6 March and 6 April 1961 he took photographs which, he claimed, showed two faint 'clouds' moving in the same orbit as the Moon, and presumably made up of meteoric material. The search had not been made at random; Kordylewski had been concentrating on these special areas, because of the connection with the famous 'three-body' problem. To explain this, we must look at those asteroids known as the Trojans.

The Trojans lie far beyond the main asteroid belt, and move in the same orbit as Jupiter, at a mean distance of about 483 million miles from the Sun. There are two groups of Trojans, one 60 degrees ahead of Jupiter and the other 60 degrees behind. They do not keep together in clumps, and may be spread out over many millions of miles, but the 60-degree points represent their mean positions, and there is no fear that any of them will wander dangerously close to Jupiter. In the diagram (fig. 1), A represents Jupiter and B and C the two Trojan groups, with S marking the Sun.

Kordylewski suggested that the same could be true of his 'clouds'. S would now stand for the Earth, with the 'clouds' at B and C. The cloud at B would keep 60 degrees behind the Moon, and the cloud at C 60 degrees ahead.

Up to now there has been no definite confirmation. Claims have been made that they were recorded in 1976 and again in 1991, but even if the clouds exist they can be nothing more than loose collections of tiny particles. No doubt we will eventually be able to decide whether they are real or not.

En passant, it is worth referring to a much older suggestion – dating back for centuries – that there may be a second Earth satellite lying directly behind the Moon, so that it is permanently hidden from us. This is an attractive idea, but it is completely unsound. The perturbations produced by other bodies in the Solar System would soon destroy the exact lining-up, and the satellite would come into view. (For the same reason, there is no possibility of an extra planet moving in the same path as the Earth and keeping behind the Sun.) We must assume that the Moon is the Earth's only natural companion, though by now there are plenty of artificial ones.

Finally, what are the chances of a satellite of the Moon itself? Ninety years ago W. H. Pickering made a careful search for one. He failed completely, and later photographic surveys by Clyde Tombaugh were unfruitful. Tombaugh concluded that there can be no natural lunar satellite with a diameter of more than about fifteen feet.

I fear that this has been another digression, but the whole question is of tremendous interest, and at least we may hope that further studies from the Moon itself will help us in clearing up some of the puzzles about the history of the Earth–Moon system.

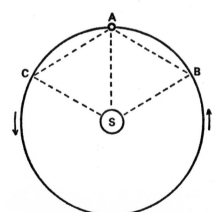

Fig. 1. Position of the Trojans relative to Jupiter, and also Kordylewski's reported objects relative to the Moon.

5

The Movements of the Moon

'The Moon moves round the Earth.' This is the bald statement given in countless books, and it is good enough for most purposes, even though mathematicians will need to qualify it. There has never been any doubt that the Moon is our companion, and that it is the nearest natural body in the sky. The Greeks knew this; so, presumably, did many of their predecessors. Where they went wrong was in supposing that the Sun, planets and stars must also circle the Earth.

Obviously the most noticeable thing about the Moon is that it seems to change in shape. Sometimes it shows up as a crescent, sometimes as a half-disk and sometimes as full, while there are periods when it cannot be seen at all. Even today there are some people who have strange ideas about these 'phases', but there is nothing in the least mysterious about them. To make things as straightforward as possible, let us imagine that the Moon is travelling round the Earth once a month in a perfectly circular orbit. This is shown in the diagram (fig. 2), which, like those following, is wildly out of scale.

The Moon has no light of its own, as the Greeks knew quite well. It shines only by reflected sunlight, and clearly the Sun can illuminate only half of the Moon at any one time. In the diagram, the unlighted and therefore non-luminous hemisphere is blackened, while the shining or day-hemisphere is left white; E represents the Earth, S the Sun, and M1, M2, M3 and M4 the Moon at various positions in its orbit.

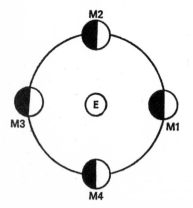

Fig. 2. Phases of the Moon

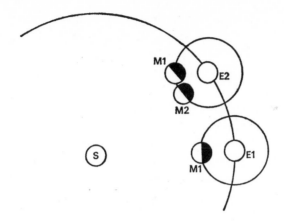

Fig. 3. The Lunation

Look first at M1. At this moment the Earth, Moon and Sun are in more or less a straight line, with the Moon in the mid position. The lighted half is (as always) turned toward the Sun, which means that the dark half is facing the Earth, and the Moon is new. People often speak of the thin crescent moon as being 'new', but this is scientifically wrong; the true new moon cannot be seen at all.

From M1, the Moon moves along in the direction of M2. Gradually a little of the day-hemisphere begins to show, and the familiar crescent makes its appearance in the evening sky; this is the best time to see the earthshine, which makes the night-hemisphere dimly luminous. By the time M2 is reached, half of the day-hemisphere is visible from the Earth, and the Moon is said to be at First Quarter.

It may sound odd to speak of the First Quarter when the Moon appears as a half-disk, but there are good reasons for it. The Moon has completed one-quarter of its journey, reckoning from new to new; also, we are seeing one-quarter of the total surface.

From M2 the Moon moves on toward M3, and more and more of the day-side comes into view. The Moon is 'gibbous', i.e. between half and full. At M3 the whole of the sunlit hemisphere faces us, and the Moon appears as a complete disk. Once again Earth, Moon and Sun are more or less lined up, but this time the Earth is in the mid position.

As the Moon continues on its way toward M4, the day-side begins to turn away from us again. Passing through the gibbous stage, the Moon has become a half-disk by the time it arrives at M4 – the phase known as Last Quarter – and it again approaches the Sun's line of sight, becoming a narrowing crescent and finally disappearing into the morning dawn. After 29½ days it has come back to M1, and is again new.

There is an obvious discrepancy here. The Moon takes only 27.3 days to go once round the Earth; why then is the interval between successive new moons over two days longer? The reason is that the Earth itself is moving round the Sun – see fig. 3, where we have the Sun at S, the Moon at M, and the Earth in two different positions, E1 and E2. When the Earth is at E1 and the Moon at M1, the Moon is new. After 27.3 days it has completed one circuit, and has arrived back at M1; but meanwhile the Earth has moved on to E2, and the Moon must travel further along its orbit, to M2, before the three bodies are properly lined up again. The extra time taken to cover the distance between M1 and M2 is just over two days, which accounts for the discrepancy.

There are technical terms for these periods. The 27.3-day revolution time is known as the Moon's sidereal period, while the interval between one new moon and the next is the 'lunation' or synodic month.

The next correction has to do with the shape of the Moon's path, which is definitely not circular. The Greeks were basically right in saying that the Moon moves round the Earth, but in almost every other respect they were hopelessly wrong. The root cause of the trouble was their utter faith in circular orbits.

Astronomers of ancient times, including Ptolemy, held that the circle is the 'perfect' form, and this meant that all the celestial bodies would have to have circular paths, since nothing short of absolute perfection could be allowed in the heavens. Unfortunately for this theory, it was obvious from the outset that the Moon does not move in a completely regular way, and neither does it always look the same size, so that its distance from the Earth must vary. One way round the difficulty would be to assume that the Earth lies some distance from the centre of the Moon's circular orbit, but even this would not account for what is actually observed. Ptolemy, an excellent mathematician, decided that the Moon must move in a small circle or 'epicycle', the centre of which – the 'deferent' – itself moved round the Earth in a perfect circle (fig. 4). As more and more problems arose the number of epicycles had to be increased, so that the whole scheme became hopelessly involved and artificial.

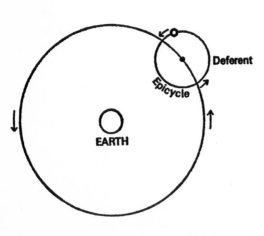

Fig. 4. Epicycles

After Ptolemy's time there was a long period of stagnation, and what we may call the modern era of astronomy dates only from the publication of Copernicus' great book in 1543. But though Copernicus was on the right track, he too was wedded to the notion of perfectly circular orbits, and he was even reduced to bringing back epicycles. The final solution was left to Johannes Kepler, whose first Law, published in 1609, stated that the

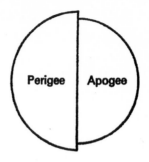

Fig. 5. Apparent size of the Moon at perigee and at apogee

planets move round the Sun in ellipses. The same is true of the Moon; it moves in an elliptical orbit, with the Earth in one of the foci.*

This explains the variation in the Moon's apparent size (fig. 5). At its closest to us, or 'perigee', the distance from the centre of the Earth is 221,500 miles; at its furthest, or 'apogee', the Moon recedes to 252,700 miles, giving a mean of rather less than a quarter of a million miles. The changes in distance are quite considerable, and the Moon's apparent diameter at apogee is only $\frac{9}{10}$ of the value at perigee. The difference is not marked enough to be noticeable with the naked eye, but it is easy to measure. On the other hand it would be most misleading to think of the lunar orbit as being highly eccentric. If it were drawn to a scale of, say, 3 inches in diameter, so that it could be fitted on to a page of this book, it would look circular unless carefully measured.

There was an interesting coincidence on 22 December 1999. The Moon was full, and was at perigee, so that it appeared as large as it ever can; the Earth–Moon system was at its closest to the Sun, so that the Moon was as brightly lit as can ever happen. Daily papers swooped on the story, claiming the Moon would be dazzlingly bright. In fact the sky was clear over my observatory in Sussex, and the full moon was brilliant – but I confess that I would not have realized that there was anything unusual about it, had I not known.

Our next correction is of a different type, and brings me back to my contention that the Earth–Moon system should be classed as a double planet rather than as a planet and a satellite. Technically, it is not correct to say that the Moon moves round the Earth. What happens is that both bodies move round their common centre of gravity.

---

* The accepted way of drawing an ellipse is to fix two pins in a board, an inch or two apart, and fasten them to the ends of a length of cotton, leaving a certain amount of slack. Then draw the cotton tight, and trace a curve, keeping the cotton tight all the time. The result will be an ellipse, with the pins marking the foci. In practice, of course, what always happens is that the thread breaks and the pins fall out.

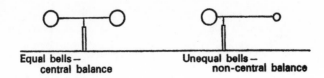

Equal bells —
central balance

Unequal bells —
non-central balance

Fig. 6. Centre of Gravity Demonstration

Consider an ordinary gymnasium dumbbell (fig. 6). Balance it on a post by its joining arm, and twist it; both bells will revolve round the centre of gravity of the system, i.e. the point where the arm is supported. Ordinarily this point will be in the middle of the arm, since the bells will be equal in weight. If one bell is heavier than the other, the balancing-point will be closer to the heavier bell; and the greater the difference in weight, the greater will be the distance of the supporting point from the centre of the arm.

The same holds good for the Earth and Moon, which may be compared with the two bells, with the force of gravity taking the place of the joining arm. The Earth has 81 times the mass of the Moon, and so the centre of gravity is shifted earthward — so far, in fact, that it lies inside the terrestrial globe, though some way from the middle of the Earth. It is around this balancing-point, or barycentre, that the two bodies are revolving.

Even now there is yet another correction to be made to our original diagram, which showed the Moon moving round the Earth in a conveniently circular orbit. This new complication arises because, strange though it may seem, the Sun's pull upon the Moon is more than twice as powerful as the Earth's.

The Earth is moving round the Sun at a mean velocity of 18½ miles per second, or roughly 66,000 mph. Relative to the Earth, the Moon's orbital velocity is only about one-third of a mile per second, so that it too is moving

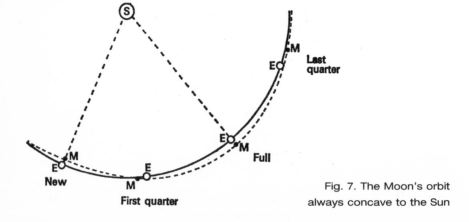

Fig. 7. The Moon's orbit
always concave to the Sun

at approximately 66,000 mph, relative to the Sun; and to an observer in space the Moon must appear as a normal planet, travelling in an elliptical orbit with the Sun in one of the foci. The path of the Moon is always concave to the Sun, as shown in fig. 7.

Although the Sun's pull is so strong, there is no chance that the Moon will part company with the Earth and move away on its own. This is because the Sun attracts the Earth and Moon almost equally. The Moon is the more strongly pulled when it lies between the Sun and the Earth, around the time of new moon, and less strongly when it is on the far side, and is full; but the force on the two bodies is always much the same.

Next, let us turn to eclipses, which have been carefully studied for many centuries, and which caused a great deal of alarm and despondence before astronomers found out why they happen.

Although the Sun is so much larger than the Moon, it is also so much further away that it appears almost exactly the same size in the sky; in each case the angular diameter is about half a degree. Consequently, when the Earth, Moon and Sun move into a line, with the Moon in the middle, the lunar disk blots out the Sun, and we have a solar eclipse. If the Moon's orbit were really as simple as shown in the first diagram, there would be a solar eclipse at every new moon. This does not happen, because the Moon's path is tilted at an angle of approximately 5 degrees to the plane of the orbit of the Earth.

The apparent yearly path of the Sun among the stars marks what we term the ecliptic, and can be calculated very accurately, even though the stars cannot be seen with the naked eye when the Sun is above the horizon (except during the fleeting moments of a total solar eclipse). The monthly path of the Moon across the sky is easy to chart, and proves to be inclined relative to the ecliptic. One way to show this is to compare the orbits with two hoops (fig. 8), hinged along a diameter and placed at an angle to each other, as shown here. The tilted hoop will lie half above and half below its companion, and this is also the case with the Moon's apparent path compared with that of the Sun. The two points where the 'hoops' cross are termed the nodes.

Unless new moon falls near a node, there can be no solar eclipse – and even if an eclipse takes place, it need not be total. Also, the Moon's shadow is only

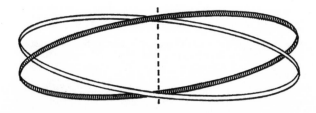

Fig. 8. Two inclined hoops

just long enough to touch the Earth, and the observer has to be in exactly the right place at exactly the right time. The last total solar eclipse visible in England was that of 1999, and we must wait until 2090 for the next, though other parts of the world are more favoured. Incidentally, no total solar eclipse can last for as long as eight minutes, and most are a great deal shorter. I once went to Siberia to see a total eclipse which lasted for a mere 37 seconds.

Lunar eclipses are quite different. They take place when the Moon passes into the shadow cast by the Earth, and the supply of direct sunlight is cut off. I will describe eclipses more fully in Chapter 13. They, too, depend upon the lining-up of the Moon, Earth and Sun, and can take place only when the Moon is full. More than a dozen total lunar eclipses will be visible from England before the end of the century.

The Moon's orbit is not absolutely unchanging for revolution after revolution. The effects due to the gravitational pulls of the Sun and Earth change, and there are other factors to be taken into account as well. The result is that the nodes shift slowly round the orbit, completing a full circuit in just over eighteen years.

Eclipses merit a chapter to themselves, so for the moment let me turn to another famous phenomenon – Harvest Moon.

Everyone knows that the Moon rises in an easterly direction and sets toward the west. This is due to the real rotation of the Earth on its axis, from west to east. The Moon is also moving in its orbit from west to east, and so it seems to travel eastward among the stars, covering about 13 degrees per day. Anyone who takes the trouble to check the position of the Moon on successive nights, using nearby stars as reference points, will soon see just what is happening.

The apparent path of the Moon in the sky is not very different from that of the Sun; the angle between the two is only 5 degrees, which is not very much even though it is sufficient to prevent eclipses from happening every month.

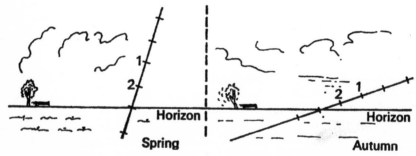

Fig. 9. Harvest Moon

When full, the Moon is opposite to the Sun in the sky, and to observers in the northern hemisphere of the Earth it lies due south at midnight.

In fig. 9, the angle of the ecliptic compared with the horizon is shown for spring and for autumn. (Again I am speaking from the viewpoint of northerners; if you happen to live in New Zealand or Australia, everything is reversed.) In spring, around March, the angle is at its steepest. In 24 hours the Moon moves from position 1 to position 2, and obviously the 'retardation' – that is to say, the difference in rising-time from one night to the next – is considerable. The situation in autumn is different. The angle is much shallower, and although the Moon moves against the starry background by the same amount – in other words, the distance between 1 and 2 is the same in each diagram – the retardation will be less. In September, the retardation may be reduced to about a quarter of an hour, no matter where you happen to be on the surface of the Earth.

It is often said that in September the full moon rises at the same time on several successive evenings. This is not true, and the retardation is always appreciable, but it is noticeably less than at other times of the year. The September full moon is called Harvest Moon, because farmers used to find it very useful as a source of extra light at a particularly busy time. The following full moon (Hunter's Moon) behaves in much the same way, though the retardation is greater.

Other full moons have old nicknames of their own. So far as I know, the full list is as follows.

January	Winter Moon, Wolf Moon
February	Snow Moon, Hunger Moon
March	Lenten Moon, Crow Moon
April	Egg Moon, Planter's Moon
May	Milk Moon, Flower Moon
June	Rose Moon, Strawberry Moon
July	Thunder Moon, Hay Moon
August	Green Corn Moon, Grain Moon
September	Harvest Moon, Fruit Moon
October	Hunter's Moon, Falling Leaves Moon
November	Freezing Moon, Frosty Moon
December	Christmas Moon, Long Night Moon

Of all these names, Harvest Moon and Hunter's Moon are the only two in common use. It has also been claimed that they look particularly big. This is quite wrong – they look no larger than any other full moons – but leads me

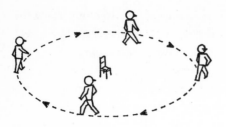

Fig. 10. Chair
Demonstration

on to the fascinating problem of the Moon Illusion. First, however, I must say a little more about the way in which the Moon rotates on its axis.

I have already touched upon the subject, when talking about the tides raised by the Earth in the viscous body of the young moon. As we have noted, the Moon's rotation was strongly braked, until relative to the Earth (though not relative to the Sun or the stars) it had stopped altogether. For thousands of millions of years now, the Moon has kept the same hemisphere turned toward us. The axial rotation period and the sidereal period are the same: 27.3 days. The Moon has what is called captured or synchronous rotation, and there is nothing coincidental or mysterious about it.

En passant, there are other bodies in the Solar System which behave in the same way. So far as we know, all the large satellites of the major planets have captured rotations. For instance Io, innermost of the principal satellites of Jupiter, has an orbital period of 1.8 Earth-days, and its axial rotation period is the same; with Iapetus, outermost of the senior attendants of Saturn, the period is as much as 79 Earth-days, and so on. The same is true for Phobos and Deimos, the midget satellites of Mars. It used to be thought that the planet Mercury, with its orbital period of 88 days, behaved similarly – in which case one hemisphere would have been permanently sunlit and the other plunged into eternal night. Science fiction writers made great play of this odd state of affairs, but recent investigations have shown that Mercury's rotation is not captured; it amounts to only 58 days, so that all parts of the planet are in sunshine at some time or other.

It has often been claimed that the Moon does not spin at all, and over the years I have had many letters about it. I have never understood why, because a very simple experiment (fig. 10) can show that the idea is absurd; we would see all parts of a non-rotating Moon. Put a chair in the middle of the room to represent the Earth, and imagine that your head represents the Moon. Stand behind the chair, a foot or two away from it, and fix your eyes upon some object beyond, such as a picture on the wall. Now walk in a circle round the chair, keeping your eyes fixed on the picture all the time. When you have completed half your circuit, you will find that the picture is in front of you and the chair behind, so that your back, not your face, is pointing chairward. To keep your face turned toward the chair all the time, you must turn as you walk; in other words you must rotate upon your axis, completing one revolution for each trip round the chair.

In fact, the Moon keeps the same hemisphere turned toward the Earth, but not toward the Sun. From Earth we can never see the 'other side' of the Moon, but conditions of day and night there are the same as on the familiar hemisphere.

Look at the Moon, even with the naked eye, and you will see obvious features; the dark plains are striking, and any pair of binoculars will show vast numbers of craters. The positions of these features on the disk are always much the same. For instance the well-marked plain which we call the Mare Crisium is always to the upper right as seen with the naked eye (as seen from the northern hemisphere), while the dark-floored crater Grimaldi is to the mid-left, and the 90-mile walled plain Ptolemæus almost in the centre. From Earth you will never see these features, or any others on the Moon, in different positions. Once you have recognized the various formations, you will find no difficulty in identifying them again, because their positions are unaltered – or nearly so.

I say 'nearly so' because there are effects known as librations which do cause slight modifications, and from Earth it is possible to examine a grand total of 59 per cent of the Moon's surface instead of exactly 50 per cent, though needless to say we can never see more than 50 per cent at any one moment. Only 41 per cent of the surface is permanently turned away from us, and therefore remained unknown until the Russians sent their first circumlunar probe on its epic journey in October 1959.

The most important libration – the libration in longitude – is due to the fact that the Moon's path round the Earth (or, to be precise, round the barycentre) is not circular, but elliptical. Kepler's Second Law, published in 1609 together with his First, states that the velocity of a planet round the Sun, or a satellite round its primary, depends upon its distance. The rule may be summed up neatly by the phrase 'the nearer, the faster'. Thus Mercury, at a mean distance of 36 million miles from the Sun, moves much more quickly in its orbit than the Earth, at 93 million miles.

The Law is equally valid for the Moon, where the distance from Earth ranges between about 221,000 miles at perigee to about 253,000 miles at apogee. When near perigee it is moving at its fastest, while by the time it reaches apogee it has slowed down. In fact, the Moon's velocity in orbit is not constant, whereas the rate of axial rotation does not change at all – a situation which has far-reaching results.

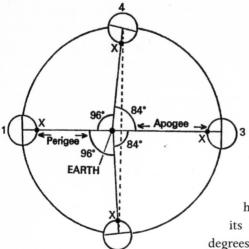

Fig. 11. Libration in longitude

The next diagram (fig. 11) should make things clear. Begin with the Moon at perigee, in position 1, and take X as marking the centre of the disk as seen from Earth. After a quarter of its journey, the Moon has reached position 2; but since it has travelled from perigee it has moved slightly quicker than its mean rate, and has covered 96 degrees instead of only 90. As seen from Earth, position X lies slightly east of the apparent centre of the disk, and a small portion of the 'far side' has come into view in the west, so that in fact we are seeing a short way beyond the mean western limb.

After a further quarter-month the Moon has reached position 3. It is now at its apogee, and point X is again central. A further 84 degrees is covered between positions 3 and 4, and X is displaced towards the west, so that an area beyond the mean, eastern limb is uncovered. At the end of one revolution the Moon has arrived back at 1, and X is once more central on the disk as seen from Earth.

Libration in longitude means, then, that the Moon seems to wobble slightly to and fro, allowing us to peer alternately beyond the mean east and west limbs of the Earth-turned hemisphere. There is also a libration in latitude; because the Moon's orbit is perceptibly tilted, we can sometimes see for some distance beyond the mean northern or southern limb. Finally there is a diurnal or daily libration, due to the axial rotation of the Earth itself; at moonrise an observer at A (fig. 12) can see slightly beyond the mean western limb, and at moonset slightly beyond the mean eastern limb.

The sum total of all these effects is that at one time or another we can study 59 per cent of the surface, but the so-called libration zones, i.e. those which are periodically brought into and out of view, are difficult to map from Earth, because all the features are seen at inconvenient angles and are badly foreshortened. Before the Space Age our charts of these regions were decidedly uncertain – as I well know, since I spent over twenty-five years in trying to map them. Luna 3, the Russian probe of 1959, marked the beginning of a new era, since it went right round the Moon and sent back the first pictures of the far side. As expected, the newly-explored regions of the Moon were just as barren and just as crater-scarred as the regions we have always known, though it is true that the two hemispheres are not exactly alike. In particular, there are no major 'seas' on the far side. Now, of course, we have very detailed maps of the entire lunar surface.

This seems the right moment to refer briefly to another peculiarity of the lunar motion: the secular acceleration, or apparent speeding-up, of the Moon on its orbit. If we take the position of the Moon as measured centuries ago, and then predict the present position by adding on the correct number of revolutions, the results will not agree; the Moon will have moved too far – or, to all appearances, too quickly.

Fortunately this can be checked, because astronomers of ancient times have left us useful eclipse records. A total eclipse of the Sun can only happen when the Moon is new, and therefore the moment of totality is also the exact time of the new moon. Eclipse records date back for thousands of years, and it was by comparing these records with modern measurements of the Moon's position that the apparent speeding-up was discovered. It is due partly to slight changes in the shape of the Earth's orbit, due mainly to Venus and Mars, and partly to the tidal effects which I described in the last chapter. The effect is very slight, but over many hundreds of years it mounts up sufficiently to become appreciable. I will say more about it later.

I appreciate that this description of the Moon's movements is very over-simplified, but I hope that the general picture is clear enough. Certainly we have learned a great deal since our Stone Age ancestors looked up at the Moon and wondered why it showed its monthly changes in form.

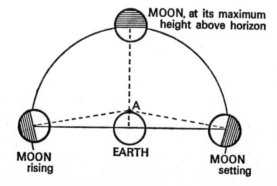

Fig. 12. Diurnal Libration

MOON, at its maximum height above horizon

A

MOON rising

EARTH

MOON setting

6

The Moon and the Earth

Unimportant though the Moon may be in the Solar System as a whole, it affects us on Earth more than any other celestial body apart from the Sun. Without the Moon, life would seem strange indeed. We could manage without its light – as we have to do for part of every month – but what about the tides?

I happen to live on the coast, within sound of the sea, so that I always know whether the tide is in or out; but there can be nobody who is not familiar with the ebb and flow of the waters, and it is obvious that the tides are associated with the Moon. The Sun is concerned too; but for the moment let us consider only the Moon – because the theory of tides is much more complicated than might be thought, and it is best to separate the various factors as much as possible.

As a start, imagine that the whole Earth is covered with a shallow, uniform ocean, and that both the Earth and the Moon are standing still. In fig. 13 (as usual, hopelessly out of scale) we have a high tide at point A, and another high tide at point B, with low tides at C and D. Basically, the Moon's gravitational pull is heaping the water up at A, where the force is strongest. Of course the pull of gravity affects the solid land also, but land is more difficult to move about than water, so that the heaping-up is less.

This is all very well, but at first sight it is not so easy to see why there should be a second high tide at B, on the far side of the Earth. It is slightly misleading to say, as many books do, that the solid globe is being simply 'pulled away'

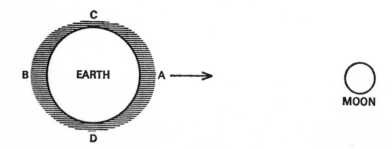

Fig. 13. Theory of the Tides

Fig. 14. Diagram to show the 'diurnal inequality' in the tides

from the water, so for a moment let us assume that we have a situation in which the Earth and the Moon are alone in the universe, and are falling toward each other because of their mutual gravitational pulls. Taking matters a step further, imagine that it is the Moon which is standing still, and the Earth which is being drawn toward it. Point A, which is closest to the Moon, will be subject to a greater accelerating force than the average, and so the water will bunch up around it, though the pull of the Earth will prevent the water from leaving the solid surface. The result will be a high tide. At B, the reverse will happen. The acceleration will be less than the average, and so the water in that region will tend to get 'left behind', so bulging away and causing a similar high tide.

Next, assume that the Earth and the Moon are keeping the same distance from each other, but that the Earth is spinning round once in 24 hours. Obviously, the water-heaps – that is to say, the high tides – will not spin with it; they will keep 'under' the Moon. Therefore, each bulge will seem to sweep right round the Earth once in 24 hours, and every region will have two high tides and two low tides per day.

Now we can start to introduce a few of the many complications. First, the Moon is not stationary; it is moving along in its orbit so that the water-heaps shift slowly as they follow the Moon around. On average, the high tide at any particular place will be 50 minutes later each day. Neither will the two daily high tides be equal. Consider a point C (fig. 14), which has a high tide. Twelve hours later it will have moved round to C^1, and will have another high tide – but this high tide will not be so great as the first, because the Earth's axis, AX, is tilted with respect to the path of the Moon. If the original tide is represented by CD, the second will be given by C^1D^1, which is decidedly less. This difference is known as the 'diurnal inequality' of the tides.

But, of course, the Earth is not surrounded by a shallow uniform water-shell. The seas are of various shapes and depths, and local effects are all-important. At Southampton, for instance, two high tides occur in succession as the rising water comes up the two narrow straits separating the mainland from the Isle of Wight, first up the Solent and then up Spithead. In the Bay of Fundy in Nova Scotia, and also along some coastal areas in South America, the tidal range is 50 feet or so, while in other places it is less than 12 inches. Actually, the forces involved are surprisingly small, and in mid-ocean the tides do not exceed a couple of feet. The majestic tides of the Bay of Fundy class are due to this deep-ocean 'wave' washing up on the shallow shelf of the coast, or being concentrated by converging coastlines.

Also, the waters take some time to heap up, and maximum tide does not occur directly under the Moon. There is an appreciable lag, and the highest tide follows the Moon after an interval which varies according to local conditions. Naturally the lag is usually greatest for shallow coastal areas.

The next complication involves the changing distance of the Moon. Near perigee, when the Moon is at its closest to the Earth, the pull is stronger than at the time of apogee, and the tides are higher; in fact the difference amounts to over 20 per cent, though again it is impossible to give a consistent figure because of the different conditions over various regions of the Earth.

Next we come to the Sun – and here we are faced with a situation which seems curious at first sight. The gravitational pull of the Sun on the Earth is much greater than the Moon's, but the Sun is of course much further away, and so far as the tides are concerned what matters is the difference of the pull at the centre of the Earth and at the point directly under the Sun. Therefore, solar tides are less than half as powerful as lunar ones; in this particular tug-of-war the Moon wins easily. What we have to do is to consider the two tide-raising bodies together.

At new or full moon – the positions when the Moon is said to be at 'syzygy' – the Sun and the Moon are pulling in the same direction (or in exactly opposite directions, which comes to the same thing) and produce the strong tides known, rather misleadingly, as spring tides. The much lower neap tides occur when the Moon is at 'quadrature', i.e. First or Last Quarter, so that its

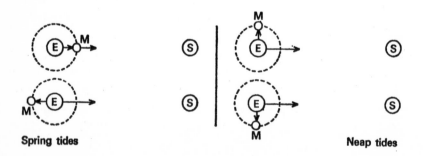

Spring tides   Neap tides

Fig. 15. Relative positions of Sun and Moon affecting tides

gravitational pull works against that of the Sun. This may sound rather involved, but I hope that the diagram (fig. 15) will make the situation clear.

The Moon pulls upon the solid body of the Earth as well as upon the oceans, and land tides are easy to measure with suitable instruments. A rather odd fact has emerged from studies of them. Although the body of the Earth behaves as though it were more rigid than steel, it also proves to be perfectly elastic; after being distorted by the tidal pull it returns to its original shape without any appreciable delay, rather in the manner of an elastic band which is first stretched and then let go. However, land tides amount to only about $4\frac{1}{2}$ inches instead of many feet, and in everyday life they are too minor to be noticed at all.

There are tides in the atmosphere, too, and it has been suggested that these may have some effect upon long-distance radio reception, which is possible because of reflecting layers in the ionosphere, tens of miles above ground level. Yet if these effects exist at all – and there is no firm evidence that they do – they are very slight indeed.

Parts of the Earth's interior are liquid, so that tides must be produced there also, and it has even been suggested that the tides inside the Earth may have some connection with earthquake shocks. In 1938 C. Davison, known for his studies of earth tremors, proposed a link between earthquakes and lunar phases, but this seems illogical, because the phases are not directly linked with the Moon's changing distance from the Earth, and the Moon is not necessarily near perigee when it is full. (For example, in May 2000 the Moon was full on the 18th of the month, but at perigee on the 6th.) I once analysed all earthquake records over a twenty-year period and tried to correlate them with the movements of the Moon, but the results were absolutely negative.

Incidentally, it is worth commenting briefly upon another idea which has been widely publicized of late. In the spring of 2000 it so happened that the main planets were roughly lined up, so that they were pulling in the same sense – and it was suggested that this might cause earthquakes and storms. In fact, the combined effects of the pulls of all the planets could raise the tides by no more than $\frac{1}{25}$ of a millimetre, which is not very much! Planetary alignments of this sort are not unusual, but can have no measurable effect upon the Earth. If an earthquake happens to occur at the time of such an alignment, it is due to sheer chance. A search for coincidences will nearly always reveal them. (I once published a short paper, linking the positions of the planets with radio reception, the fluctuations of the remote variable star Delta Cephei, and the frequency of matinée performances at the Folies Bergère in Paris. Delta Cephei is over 1,300 light-years away, so that it does not seem likely that its behaviour can have much to do with us; I cannot comment upon the second case, because I have not been to the Folies Bergère since I last attended a scientific conference in Paris some time ago.) Coincidence-hunting can be amusing, though completely useless. I have no doubt that by selecting suitable criteria it would be possible to draw graphs showing links between the periods of the minor planets, the price of bananas, and the number of goals scored by the Manchester United football club. Astrologers are particularly good at this sort of thing.

Neither is there any valid connection between the Moon and weather. True, the weather does often change at full moon, but – after all – it changes every few days, in England at least. An old country saying tells us that 'the full moon eats up the clouds', and certainly the sky sometimes clears when the full moon rises – but at the same time the Sun must be setting, and nobody can deny that weather and cloud conditions are dependent upon the Sun.

All the same, it is worth saying a little about some of the atmospheric effects seen during moonlight, even though they belong strictly to the realm of meteorology and have no actual connection with the Moon in the astronomical sense.

Everyone knows the expression 'once in a blue moon', but how many people have really seen a blue moon? I have – once, on 26 September 1950, when I was living at East Grinstead in Sussex. Late in the evening I recorded that 'the Moon shone down from a slightly misty sky with a lovely shimmering blueness – like an electric glimmer, utterly different from anything I have seen before'. The phenomenon was certainly not confined to East Grinstead. Over a period of at least 48 hours, people in various parts of the world were startled to see not only a blue moon but also, in some places, a blue sun, and some of the Press reports were highly sensational. Yet there was no mystery about it. Giant forest fires raging in Canada had sent a tremendous amount of fine dust into the upper atmosphere, and it was this dust which caused the eerie colouring. From all accounts, the dust-pall over North America was really remarkable. At Ottawa, car headlights had to be switched on at midday, even though there was no fog, and in New York a game of baseball was played under arc-lights. Other blue moons have been seen occasionally, always for the same sort of reason.*

Haloes, or luminous rings round the Moon's disk, are comparatively common, and can be beautiful. They are not due to dust, but to moonlight shining upon ice crystals in the upper air, about 20,000 feet above the

---

* It is often said that if two full moons occur in the same calendar month, the second is a 'blue moon'. This has nothing to do with colour, and is not an old legend; it arose from a misinterpretation of the data given in an American farmers' almanac.

ground. These crystals make up the type of cloud known as cirrostratus. If the cloud is lower and denser, the Moon merely looks watery. Both watery moons and haloes are said to be forerunners of rain, and this is often true, because cirrostratus cloud is itself a common sign of approaching bad weather. Paraselenæ, or 'mock moons' – brilliant images of the Moon some way from the true disk – are also due to ice crystals, but are very rare. I have yet to see one.

When the Moon shines upon water droplets in the atmosphere it may produce a rainbow in the same way as the Sun, but lunar rainbows are relatively faint and rare, as well as being less brightly coloured, simply because the intensity of moonlight is so much less than that of sunlight; as I have already said, it would take more than half a million full moons to provide as much light as the midday Sun. Yet now and then a striking lunar rainbow is seen. Flying at about 2,000 feet above Scotland during a war-time operation on 28 March 1945, I was particularly lucky; most of the rainbow circle could be seen, and even some delicate, fugitive hues, giving a lovely effect. As I was the navigator of the aircraft I had little time to study the rainbow, but I am glad to have seen it.

The celebrated Moon Illusion, which has been the subject of a good deal of research, is not an atmospheric phenomenon at all, and in fact it still remains something of a puzzle. For some reason or other, the full moon seems to look larger when close to the horizon than when it is high in the sky. Indeed, the casual observer is apt to say that it looks twice the size, whatever the state of the sky – particularly at the time of Harvest Moon. Actually, the Moon is slightly more distant when low down than when high up. The observer is brought toward the Moon as the Earth turns so that the high Moon is a little the larger, though the difference amounts to less than 2 per cent. Why, then, should the low Moon look the bigger?

It is not really so, as measurements prove, so that some trick of the eye or the brain is responsible. (In passing, the Moon never looks so large as most people imagine; try covering it with a small coin held at arm's length, and you will see what I mean.)

The illusion has been known for many centuries. So far as I know, the first man who tried to explain it was Ptolemy, around AD 150, and his theory is certainly better than many of those put forward in modern times, though it is not likely to be the complete answer. Ptolemy pointed out that the low-down Moon is seen against a foreground of 'filled space' (trees, houses and so on), and so there are nearby objects to act as comparisons; when the Moon is well above the horizon there are no comparisons, and we look at it across 'empty space'. When low, then, the Moon will seem to be more remote than when it is high; and if the images seen in the eye are of equal size, the disk which is the further away will seem to be the larger.

Yet will the Moon really seem more distant when we look at it behind 'filled space'? One man who disagreed was George Berkeley, who in 1709 published a completely different theory. First he tried to dispose of Ptolemy's idea, by looking at the rising full moon through a tube and thereby cutting out any foreground; he claimed that the illusion was still obvious. He went on to suggest that because the low Moon is shining through a relatively thick layer of the Earth's air, it appears fainter than the high Moon (because more of the moonlight is absorbed), and it is this which makes the low Moon seem both more remote and oversized. Berkeley's theory was supported in 1973 by Professor E. J. Furlong of Dublin University, though in modified form. Furlong held that a change in the Moon's colour, again due to its being seen through a thicker layer of atmosphere, is an important factor.

Some years ago, E. G. Boring at Harvard carried out experiments which indicated that the illusion is due to the mechanism of the human eye; the unconscious effort of raising the eye to look at a high-up object causes the Moon to look smaller than it really is. Boring's work was followed up in 1959 by H. Leibowitz and T. Hartman, at Wisconsin, who experimented with disks seen at eye-level and overhead. They concluded that the illusion is due to the fact that we have more visual experience with objects in the horizontal plane than in the vertical. They added that children were more affected than adults.

The next major paper on the subject, by L. Kaufman and I. Rock in 1962, was essentially a return to Ptolemy's theory. Other writers claimed that one-eyed observers were not conscious of the illusion, and there was also a suggestion that dustiness in the atmosphere near the horizon blurs the edges of the Moon and enlarges the disk.

It all seemed very uncertain, and in 1974 I presented a television programme about it. With me was Professor Richard Gregory, who is an acknowledged expert upon illusions in general (as well as on many other subjects!). The first thing we tried to do was to test the Berkeley–Furlong theory. We fixed up a white disk, and illuminated it from behind; then we observed it from a distance, dimming it and changing its colour by means of filters put in front of it. There was no perceptible difference in apparent size, so we turned to an experiment entirely our own – though I must stress that Professor Gregory designed it, and that I was merely the operator.

We went out on to the Selsey coast at the dead of night, armed with a system of movable mirrors. The full moon was high up, and the sky was clear. We could produce an artificial moon whose size could be altered by means of an iris, and by swinging the mirrors we could alter the apparent altitude of the image of the real Moon. The idea was to test not the real sizes of the disks, but their apparent relative sizes, which is not the same thing. While the senior experimenter operated the mirrors and lights, I stood well back and compared the two disks, estimating them as accurately as I could and using a special measuring device. First the real Moon was 'brought down' until it was side by side with the artificial one, and the iris was adjusted until, to me, the two disks were equal. Then the real Moon was raised – and it seemed to shrink; only when the artificial image had been reduced by about 10 per cent did the two seem equal again.

We had also persuaded a member of our team to cover up one eye for several hours beforehand, and found that the illusion was still there. To test the behaviour of my eyes when looking at objects from unfamiliar angles, I observed both the low Moon and the high Moon while standing on my head, thereby causing onlookers to class me as insane. Again there was no difference in the illusion. We finally concluded that Ptolemy had been on the right track, but that foreground, eye and brain were all involved. Since the illusion is essentially a physiological one, and is not due to the Moon itself, further discussion of it here would, I fear, be out of place.

Can there be any connection between the Moon and human behaviour? 'Lunar' and 'lunacy' have long been linked, and it is widely believed that people who are mentally unstable are at their very worst near the time of the full moon. I can make no contribution here, because I know nothing about medicine – but I once carried out a survey by asking for the opinions of experts; opinions were sharply divided. Of course, aquatic creatures do regulate their behaviour according to the phases of the Moon, but this is due to the tides. Any links between the Moon and land plants are, to put it mildly, tentative.

All in all, it seems that the Moon's influence upon us is limited mainly to the movement of the Earth and the regulation of the tides. Once we try to take matters any further, we are entering the realm of speculation.

It is widely believed that some meteorites found on Earth have come from the Moon – for example the stones MAC 88104 and 88105, found in Antarctica in 1990. They may have been blasted away from the Moon by a huge impact, and were lying in the frozen Antarctic wastes for at least 30,000 years before being identified. Over a dozen 'lunar meteorites' have been listed, and though we cannot be sure that they come from the Moon, the evidence is fairly strong. There are also the small, dark, glassy objects known as tektites, found in a few localized areas. They were once thought to be lunar, and to have been sent out from lunar volcanoes, but it now seems fairly certain that they are of terrestrial origin. Even the largest of them is smaller than a hen's egg.

Finally, suppose that the Moon had never been formed – how would the Earth differ from the world we know? In fact, the differences would be much more pronounced than might be expected, quite apart from the lack of moonlight. The Moon has played a major rôle in slowing down the Earth's axial rotation; without it, the 'day' would be much shorter than it actually is. Quick rotation leads to strong windspeeds, as on Jupiter and Saturn. On a moonless Earth there would be frequent gales, and considerable erosion, so that there would be much less surface relief and certainly no mountains as high as the Himalayas. Tides would be purely solar and relatively mild, so that life might well have evolved in a completely different way. Even more importantly, it seems that the Moon has a stabilising effect upon the Earth, and keeps our axis inclined more or less at its present angle of $23\frac{1}{2}$ degrees to the perpendicular to the orbit. On Mars, where the two satellites (Phobos and Deimos) are too small to have any measurable influence, the axial tilt ranges between 35 degrees and only 15 degrees over a cycle of about 50,000 years. If the Earth were subject to the same changes, the seasons would be very variable, probably with adverse effects upon life. It has even been suggested that we owe our very existence to the Moon. This may be too extreme, but certainly there is no doubt that we have every reason to be grateful to our faithful companion in space.

7

Observers of the Moon

The 'seas' of the Moon are easy to see with the naked eye, and their shapes are well defined. Therefore, there should be no problem in drawing an outline map. It is usually said that the first drawing to show anything recognizable was made by Leonardo da Vinci, around 1505, and although the sketch is fragmentary it does seem to indicate the positions of some of the main features, including the Mare Crisium; but we now have evidence that the earliest map goes back much further than that – in fact, to around the year 3000 BC.

Ireland is noted for its prehistoric monuments. The passage-tomb of Newgrange is a famous tourist attraction, and in the same area – the Boyne valley – are two others, Knowth and Dowth. Knowth consists of a large mound surrounded by seventeen others of smaller size. There are elaborate rock carvings and beautifully decorated stones, and one of these appears to be a very rudimentary chart of the Moon: it was identified in 1999 by Dr Philip Stooke, of the University of Western Ontario in Canada, who is one of the world's leading lunar and planetary cartographers and is deeply interested in the history of science. Of course the map – if you can call it that – shows little more than a few arcs, but they do correspond to the main lunar maria, and it is difficult to disagree with Dr Stooke's comment that 'the people who carved this Moon map were the first scientists. They knew a great deal about the motion of the Moon; moreover it is found that at certain times of the year moonlight can shine down the eastern passage of the tomb – on to the Neolithic etching.'*

Come now to historic times, and to William Gilbert, physician of Queen Elizabeth I. He is best remembered as the pioneer investigator of magnetic phenomena, but he was interested in all branches of science, and his map of the Moon (fig. 16) was drawn around 1600. The map was not actually published until 1651, but it must have been completed before 1603, for the excellent reason that this was the year in which Gilbert died. The main dark regions are shown in recognizable form; for instance, the patch which he calls the 'Regio Magna Orientale' corresponds to our Mare Imbrium.

---

* The discovery of the map caused considerable popular interest at the time. I was vastly amused by a headline in the London *Daily Mail* for 23 April: 'How a Prehistoric Patrick Moore Mapped out the Moon!'

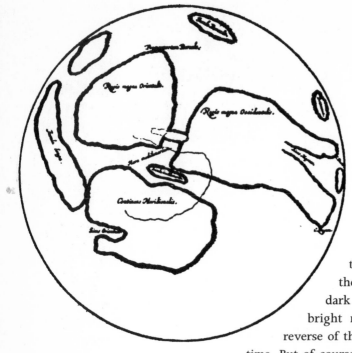

Fig. 16. William Gilbert's
naked-eye Moon map

Gilbert's map was
much the best of pre-
telescopic times, even
though he did regard the
dark areas as land and the
bright regions as seas – the
reverse of the general view at that
time. But of course he had to depend on
the naked eye alone; telescopes did not come into use for some years after he
died. (In fact I have always been surprised that the telescope was not invented
much earlier than it actually was; spectacle lenses to help people with poor
eyesight were certainly in use by 1300, and by Gilbert's time spectacles had
started to look rather like those of today.)

The first telescopes of which we have definite knowledge were made in
Holland in 1608, and the Moon was obviously one of the first celestial bodies
to be examined. Of the early observers, Galileo was by far the greatest, but he
was not the first. This honour seems to go to Thomas Harriott, one-time tutor
of Sir Walter Raleigh, who was an excellent mathematician and who obtained
a telescope from Holland soon after they were invented. I feel able to make a
comment here, because I played a rôle, albeit a very minor one, in the first
publication of Harriott's lunar map (fig. 17).

In 1965 I received a letter from Dr E. Strout, of the Institute of the
History of Science in what was then the USSR, enclosing a paper he was
writing; he had even located the map, which had apparently been drawn in
July 1609. I was at that time Director of the Lunar Section of the British
Astronomical Association, and I was able to arrange the prompt publication
of the Strout paper, plus the Harriott map, in the Association's *Journal*
(volume 75, page 102).

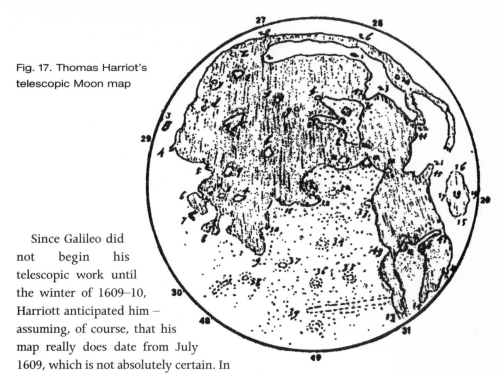

Fig. 17. Thomas Harriot's telescopic Moon map

Since Galileo did not begin his telescopic work until the winter of 1609–10, Harriott anticipated him – assuming, of course, that his map really does date from July 1609, which is not absolutely certain. In any case, the chart contains many features which are clearly recognizable, and in view of the weakness and poor definition of telescopes at that time it was a remarkably good effort – much better, candidly, than anything which Galileo produced. Unfortunately Harriott never followed it up. He died in 1621.

One of Harriott's correspondents was Sir William Lower, who lived in the Welsh village of Treventy. Lower is a shadowy, elusive figure. He was certainly expelled from Oxford University in 1591, following a celebration which was presumably anything but teetotal; from 1601 to 1611 he sat in Parliament, first as member for Bodmin and then for Lostwithiel; he was knighted in 1603, and after his marriage to one Penelope Perrot, daughter and heiress of Thomas Perrot of Treventy, he retired to the seclusion of Wales to spend much of his time in studying astronomy. Very little is known about his work, but he did leave a graphic description of the full moon as resembling a tart that his cook had made – 'there some bright stuffe, there some dark, and so confusedlie all over'. This does not sound ultra-scientific, and I have been unable to discover any observations by Lower which were of true value; but at least he looked at the Moon, and we may suppose that he went on doing so until he died in 1615.

So let us turn to Galileo, who will always be remembered as the real pioneer of telescopic observation. He began his observations in January 1610, and of course the Moon was one of his first targets. He left a good description of it:

... I distinguish two parts of it, which I call respectively the brighter and the darker. The brighter seems to surround and pervade the whole hemisphere; but the darker part, like a sort of cloud, discolours the Moon's surface and makes it appear covered with spots. Now these spots, as they are somewhat dark and of considerable size, are plain to everyone and every age has seen them, wherefore I will call them great or ancient spots, to distinguish them from other spots, smaller in size, but so thickly scattered that they sprinkle the whole surface of the Moon, but especially the brighter portion of it. These spots have never been observed by anyone before me; and from my observations of them, often repeated, I have been led to the opinion which I have expressed, namely, that I feel sure that the surface of the Moon is not perfectly smooth, free from inequalities and exactly spherical... but that, on the contrary, it is full of inequalities, uneven, full of hollows and protuberances, just like the surface of the Earth itself, which is varied everywhere by lofty mountains and deep valleys.

The smaller spots are, of course, craters. Galileo's description seems 'modern' enough, and he went on to make measurements of the heights of some of the lunar mountains. His results were rough, but they were at least of the right order of magnitude. Galileo found that the lunar mountains are much higher, relatively, than those of the Earth; he realized that the Moon is a rugged world.

Galileo made a number of lunar drawings, and on some of them it is possible to identify various features, though the accuracy is low (fig. 18). Unlike Harriott, Galileo never compiled a proper map, and it seems that after those early days he paid little further attention to the Moon's surface; he was busy with greater problems – and, of course, with his battles against the Church, which make fascinating reading but which do not concern us here. Remember, too, that Galileo's telescopes were very feeble. Even the most powerful of them magnified no more than thirty times.

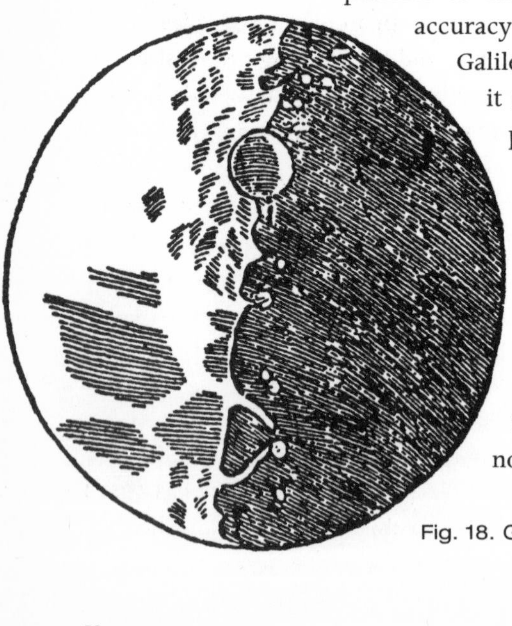

Fig. 18. Galileo's drawing of the Moon

Galileo made one more comment about the Moon in 1641, after the infamous trial at the hands of the Inquisition, and when he was old, blind, and living under strict Church supervision in his lonely villa. A scientist named Fortunio Liceti had claimed that the faint luminosity of the night side of the Moon, seen during the crescent stage, was due to the diffusion of sunlight in a lunar atmosphere. Galileo argued, quite correctly, that Leonardo da Vinci had been right in saying that it must be due to light reflected on to the Moon from the Earth.

The first stage of lunar telescopic research ended with Galileo. For the first time the mountains and craters could be clearly seen; it was still generally believed that the dark areas were watery, and there seemed no obvious reason to doubt that there could also be life. One man who certainly thought so was Johannes Kepler, who corresponded with Galileo and who drew up the famous Laws of Planetary Motion which gave the death-blow to the theory that the Sun goes round the Earth. Kepler went so far as to write a science-fiction story in which he gave a vivid description of the lunar life-forms – some of them serpent-like, others aquatic, and all of them covered with thick fur! Kepler's story, the *Somnium*, was published posthumously in 1634.

The next major advance came in 1647, and was due to Hevelius (more properly, Hewelcke), a city councillor of Danzig in Poland, now known as Gdańsk. Hevelius built an observatory on the roof of his house, and equipped it with the best instruments available at the time. His telescopes were strange and awkward; they had small object-glasses and very long focal lengths, and were incredibly cumbersome. (In some telescopes the object-glass had to be entirely separate, and fixed to a mast. A French astronomer named Auzout once designed a telescope 600 feet long, though apparently it was never built.) Yet Hevelius was a patient, skilful observer, and his map (fig. 19), just under a foot in diameter, showed many identifiable features. He also made height-measurements of some of the lunar peaks which were better than Galileo's.

Hevelius gave considerable thought to the best method of naming lunar features. He finally decided to give them terrestrial names, and this was the system followed on his map. For instance, the crater now called Copernicus was his 'Etna', while our Plato was 'the Greater Black Lake'. It was a reasonable idea, but it did not work well, and less than half a dozen of Hevelius' lunar names are still in use. Copies of the map exist, but the original copperplate of it is no longer to be found. After Hevelius' death it was apparently melted down and made into a teapot.

Next came Giovanni Battista Riccioli, an Italian Jesuit who published a lunar map in 1651. It was about the same size as Hevelius', and was based largely upon the observations made by Riccioli's pupil, Grimaldi. The map

Fig. 19. Hevelius'
lunar map

itself was fairly good, but it is remembered chiefly for its system of
nomenclature. Riccioli gave the dark plains romantic names such as the Mare
Imbrium (Sea of Showers), Mare Tranquillitatis (Sea of Tranquillity) and Sinus
Iridum (Bay of Rainbows); the craters were named after famous personalities
– usually, though not always, astronomers. The system quickly replaced that
of Hevelius, and nearly all the names he gave (more than 200 altogether) are
still in use. Later selenographers have added to the list, and the whole scheme
is so deeply rooted that it will certainly never be altered now.

Riccioli was a man of firm opinions, some of which were completely
wrong. He could never bring himself to believe that the Earth is in orbit
round the Sun, and this meant that he had little patience with Copernicus,

who had revived the Sun-centred or heliocentric system over a hundred years earlier. Alas, Copernicus had to have a crater; but Riccioli, to quote his own words, 'flung Copernicus into the Ocean of Storms', which is why the formation honouring the great Polish scientist is to be found in the midst of the vast Oceanus Procellarum. At least the crater is large and important, but Galileo was not so well treated, and was allotted a very obscure crater towards the edge of the Oceanus. On the other hand, Ptolemy was given a magnificent walled plain close to the middle of the Moon's disk, and to Tycho Brahe, the eccentric Dane, went the most prominent crater on the Moon; it lies in the southern uplands, and is the centre of the ray-system which dominates the whole lunar scene near full. Riccioli was not modest. He and Grimaldi are represented by two huge, dark-floored walled plains which are recognizable at any time when they are in sunlight.

One trouble about all this was that most of the chief craters were used up, so that later students of the Moon had to be content with second-best. Sir Isaac Newton, who was a small boy when Riccioli drew his map, is tucked away near the Moon's south pole; Johann Mädler, about whom I shall have much to say shortly, has an insignificant crater on the Mare Nectaris or Sea of Nectar, and so on. There are also some rather unexpected names. Julius Cæsar is there, presumably because he was responsible for the reform of the calendar; so are Alexander the Great and his friend Nearch; there are even a couple of Olympians, Atlas and Hercules. One crater has been given the rather startling name of Hell. This does not, however, indicate exceptional depth; it commemorates Maximilian Hell, a Hungarian astronomer of the eighteenth century. Also on the Moon we find Barrow (Isaac Barrow, Newton's tutor) and Birmingham (John Birmingham, an Irish observer who died in 1884).

At one stage, not so long ago, the whole scheme of lunar nomenclature began to get out of hand. Finally, the International Astronomical Union, the controlling body of world astronomy, standardized matters, and the situation today is much more satisfactory. Of course, it has also been necessary to name the features on the Moon's far side, never visible from Earth; but that is running ahead of our story.

Passing over a hundred years, and pausing only to mention the 21¼-inch map of the Moon drawn by Giovanni Cassini in 1680, we come to Tobias Mayer, a German astronomer whose lunar chart was published in 1775, thirteen years after his death. Mayer introduced a system of co-ordinates (the equivalents of latitude and longitude on the Moon) and produced an excellent 8-inch map which remained the best for over half a century. Unlike Hevelius, Riccioli and most of his predecessors, he drew the Moon with south at the top, which is how telescopic observers actually see it — provided, of course, that they live in the northern hemisphere of the Earth. (Astronomical telescopes give an inverted image. I will return to this problem later.)

Soon after the appearance of Mayer's map, Johann Hieronymus Schröter founded a private observatory at Lilienthal, near Bremen in Germany, and embarked upon a research programme which was really the beginning of modern-type selenography. He was never a professional astronomer. For much of his career he was chief magistrate of Lilienthal, with ample means and leisure to carry on his hobby, and he acquired several reflecting telescopes, two of which were made by no less a person than William Herschel. The largest instrument had a 19-inch mirror, and was the work of Schräder of Kiel. For thirty years he worked patiently away, drawing, measuring and charting (fig. 20). To a great extent he was breaking new ground, and it was he who, for instance, first studied the crack-like features which we now call clefts or rills. He did not actually discover them (this honour must go to the great Dutch scientist Christiaan Huygens, a century earlier), but Schröter was the first to chart them in detail.

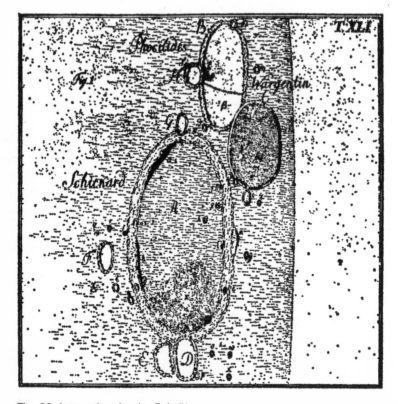

Fig. 20. Lunar drawing by Schröter

Schröter has been much maligned. It has been claimed that all his work was inaccurate, and even that his telescopes were of poor quality. I can only say that I disagree most strongly. It is true that he was not a good draughtsman; his drawings are clumsy and schematic, and the 19-inch Schräder reflector may well have been imperfect. Also, some of Schröter's ideas were very wide of the mark. He believed that he had found important structural changes on the Moon's surface, he considered that there must be a fairly dense atmosphere, and he was quite prepared to accept the idea of intelligent life there. On the other hand he was a completely honest observer, who never drew anything unless he could be quite certain in his own mind that he had seen it, while his measurements of mountain-heights were better than any previously made. As for his telescopes – well, whatever may be said about the 19-inch, there can be no doubt about the high quality of the two made by Herschel.* The more I study Schröter's work, the greater my admiration for it.

Unfortunately, despite his hundreds of drawings of various parts of the Moon, he never made a complete map, and his career came to a sad and abrupt end. In 1813, when he was approaching the age of seventy, the French, under Vandamme, occupied Bremen; Lilienthal fell into their hands, and Schröter's observatory was burned to the ground, along with all his notes, manuscripts and unpublished observations. Even the telescopes were plundered, because they were brass-tubed, and the French soldiers thought that they must be gold. The loss could never be made good, and the old astronomer had no time to begin again; he died three years later.

The mantle of Schröter fell upon three of his countrymen: Lohrmann, Beer and Mädler. All were clever draughtsmen as well as being good observers, and between them they explored every square mile of the Moon's visible surface – but they had Schröter's work to use as a basis. The credit for founding the science of precise lunar observation should go to the Lilienthal amateur above all others.

Wilhelm Lohrmann, a Dresden land surveyor, set out to compile a really large, accurate lunar map. The first sections, published in 1824, were amazingly good. Unfortunately he had completed only three more sections when his eyesight failed him, and he had to give up. He died in 1840.

Next came the superb work by a Berlin banker, Wilhelm Beer, and his friend Johann Mädler. They used the $3\frac{3}{4}$-inch refracting telescope at Beer's home, and studied the Moon intensively for more than ten years, finally producing a map which was the basis for all later studies right up to the start

---

* As an interesting historical aside, so far as we know, the only telescopes made by Herschel which were ever used for serious work were his own and Schröter's. Yet during his lifetime Herschel built, and distributed, many dozens of high-quality instruments.

of the Space Age. They followed it up with a book, *Der Mond*, which contains a full description of the whole of the visible surface. *Der Mond* appeared in 1838, and made a tremendous impact upon the astronomical world. For some strange reason it has never been translated into English. (Had I been able to read German, I would have put this right years ago.)

Oddly enough Mädler, who did most of the mapping, used the 3¾-inch telescope for almost all his lunar work. It is true that, inch for inch, a refractor is more effective than a reflector; even so, the difference between Mädler's telescope and Schröter's instruments is remarkable. No doubt Mädler's gave a sharper image, but it must have been inferior in sheer light-grasp.

Der Mond not only revolutionized selenography, but also actually held it back to some extent! Schröter had believed the Moon to be a living, changing world, but Beer and Mädler went to the other extreme, and considered that it must be absolutely dead. Their opinions naturally carried a great deal of weight. Neither of them did much more lunar work after 1840, when Mädler left Berlin to become director of the new Dorpat Observatory in Estonia (then occupied by the Russians), and nobody else seemed really inclined to follow in their footsteps. If their map were the last word on the subject, and if the Moon were a changeless world, what could be the point of observing it further?

Luckily there was one astronomer, Julius Schmidt, who did not agree. He began observing the Moon during his boyhood, and continued to do so until his death in 1884. After acting as assistant at various German observatories he was appointed to the directorship of the Athens Observatory in 1858, and it was in Greece that most of his best work was done. Schmidt not only revised and completed the map begun by Lohrmann, but also produced one of his own (fig. 21). It was 74 inches across, and stands up well when compared with modern charts. Yet before this map appeared – in 1878 – much had happened, and Schmidt was chiefly responsible. It was the 'Linné affair' which reawakened popular interest in the Moon.

At various times Lohrmann, Beer and Mädler, and Schmidt himself, had recorded a deep crater in the Mare Serenitatis or Sea of Serenity. Mädler had named it Linné in honour of Carl von Linné (Linnæus), the Swedish botanist. Then, in 1866, Schmidt announced that the crater was no longer there. It had vanished from the Moon, or altered in appearance beyond all recognition.

This was startling, to put it mildly. Could the Moon be less dead than Mädler thought? It was a revolutionary idea, and yet, coming from an observer with Schmidt's reputation, it could not be disregarded. Amateurs and professionals alike were intrigued, and once more telescopes were pointed at the lunar surface. The number of papers and discussions about this minor feature of the Moon became astronomical in every sense of the word.

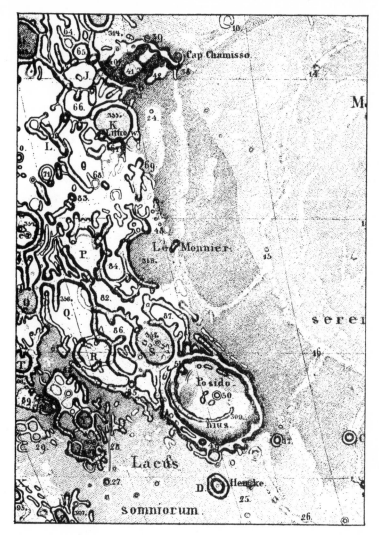

Fig. 21. A map from Dr Julius Schmidt's *Mondcharte*, published 1878

We now know that there has been no change in Linné. It is a small, normal impact crater, and there is nothing special about it. Yet the Schmidt announcement proved to be very useful indeed, and it is fair to say that 1866 marked the start of regular, careful surveying of the Moon.

Up to that time the Germans had taken the lead, but after the Linné alarm others joined in. The first important British book on the subject was the work of Edmund Nevill, who wrote under the name of Neison. *The Moon*, published in 1876, contained a map based on Mädler's, together with a description of every named formation (fig. 22). It was of tremendous value; even today copies

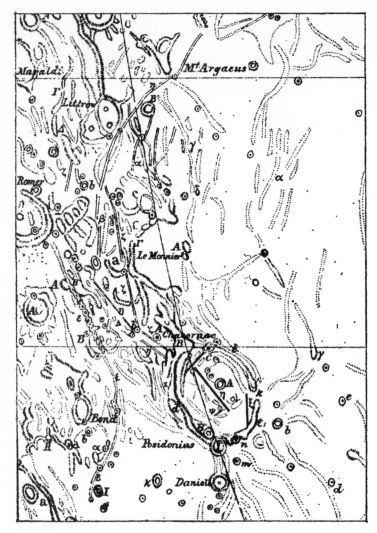

Fig. 22. The same area, taken from Edmund Neison's
The Moon, published 1876

of it can be picked up occasionally, and even when I began looking at the
Moon, around 1930, it was still regarded as a 'must' for every lunar observer.

Neison himself provides a link between the past and the near-present. He
was only twenty-five when he wrote his book, and when I first read it I
naturally thought that he must have been dead for many years. In fact he was
not; he was living at Eastbourne, only thirty miles from my old home at East
Grinstead, and I could well have met him, though to my great regret I never
did so. He died in 1938. He did very little lunar work after 1882, when he

went to South Africa to become director of the new observatory in Natal – which was unfortunately closed down in 1912 because of lack of funds.

At about the time that Neison's book was published, a new society devoted entirely to studies of the Moon was formed in England. It was called the Selenographical Society, and for ten years or so it was very active. In 1883, following the death of its president (W. R. Birt) and the resignation of its secretary (Neison), it was disbanded, but seven years later the newly-founded British Astronomical Association set up a Lunar Section and carried on the work.

Meanwhile, photography had begun to play a rôle. In 1839 François Arago, France's leading astronomer, gave a speech to the Chamber of Deputies in Paris, and made a prediction which is worth quoting: 'We may hope to make photographic maps of our satellite, which means that we will carry out one of the most lengthy, most exacting and most delicate tasks of astronomy in a few minutes.' In fact things were not so simple as Arago had expected, and it is probably true to say that lunar mapping was completed only in 1967, with the photographs sent back by the Orbiter rocket probes; but it was not long before the camera started to replace the eye in many ways.

On 23 March 1840, J. W. Draper, of New York, made a Daguerreotype of the Moon, using a long-focus 5-inch reflector. The image was a mere inch in diameter, and the exposure-time needed was as long as twenty minutes, but various light and dark areas were recorded. Ten years later J. A. Whipple, at Harvard, obtained a good series of Daguerreotypes showing the Moon at different phases, and some of these pictures were capable of being enlarged to a scale of five inches to the Moon's diameter. (I remember including some of them when I took part in arranging the Moon display at the Festival of Britain in 1951. They had also been shown at the Great Exhibition of 1851, causing a great deal of interest.)

The development of the wet collodion process helped matters considerably, and in 1852 some good results were obtained by an English amateur, Warren de la Rue. Later, using dry plates, de la Rue in England and Rutherfurd in the USA were able to take photographs which were of definite astronomical value. Others, too, began to join in, and after 1890 lunar photography became all-important.

During the last decade of the nineteenth century and the first of the twentieth, true photographic atlases began to appear: one from Paris, one from Belgium, one from the Lick Observatory in America, and so on. The two main Paris observers, Loewy and Puiseux, produced an excellent series of plates covering the entire visible face of the Moon, and subsequently all lunar mapping was photographically based. S. A. Saunder, an English amateur who was by profession a schoolmaster, drew up a list of the positions of lunar features which was of tremendous value; and in 1904 W. H. Pickering, one of the few professional astronomers to show a real interest in the Moon, issued an atlas in which he showed each region under five different conditions of illumination. But Pickering was more or less on his own; professional astronomers as a whole were not in the least interested in the lunar surface. To them, the Moon was dull and parochial. It never changed, but twentieth-century astronomy was changing very rapidly indeed, and virtually all the attention was concentrated upon the stars and the remote star-systems. Great new telescopes came into use, spearheaded by the 100-inch Hooker reflector on Mount Winson in California, but these were seldom turned moonward. The 100-inch was used to take a few pictures of lunar craters, simply to satisfy public demand, but that was all.

Amateurs had the field to themselves, and they rose nobly to the occasion. Successive directors of the British Astronomical Association's Lunar Section produced good maps: T. G. Elger in 1896, Walter Goodacre in 1930 and Percy Wilkins in 1946, but from a professional point of view the Moon was ignored. It is true that the International Astronomical Union issued a map in 1935, but it was not clear enough to be really useful, and it was defective near the Moon's limb, where the formations are so foreshortened that mapping is very difficult indeed.

In 1959 came the first professional lunar photographic atlas since Pickering's; it was compiled by G. P. Kuiper in the United States, and was based on the best photographs available. It was a major contribution, but it was still unsatisfactory. Over-enlargement led to blurring of many of the features, except near the centre of the disk, and the limb regions, alternately carried in and out of view by the effects of libration, were poorly shown.

Then came the start of the Space Age – and a complete change in outlook. Travelling to the Moon was no longer classed as science fiction; by 1960 most people were confident that the Moon was within reach. In fact, little more than a dozen years elapsed between Yuri Gagarin's initial foray into space and Neil Armstrong's first step on to the lunar surface.

So far as lunar mapping was concerned, space research methods took over. The first successful probe was Lunik 1 (or Luna 1) sent up by the Russians in January 1959. Eight months later Luna 2 crash-landed on the surface, and then, in October 1959, came the first photographic probe – Luna 3, which went right round the Moon and sent back pictures of the previously unseen far side. The next step came in 1964, when the Americans had their first major success; Ranger 7 sent back more than 4,000 pictures of the Mare Nubium before destroying itself on impact. Two more Rangers followed, and then, between October 1966 and January 1968, the five Orbiters, which surveyed the entire surface of the Moon. It was the Orbiter programme which really superseded visual mapping from the surface of the Earth. François Arago's prophecy had come true – even though it did take a little longer than he had expected when he spoke at Paris in 1840.

This is not to say that observations from Earth are no longer useful. On the contrary, they are very valuable indeed, as I hope to show later in this book; I am referring only to sheer cartography. After all, we now know the Moon's surface better than we know some areas of the Earth, such as tropical rain forests.

In December 1968 Apollo 8 went round the Moon, carrying Astronauts Lovell, Borman and Anders; for the first time, human observers were able to see the Moon from close range. The Apollo programme ended in December 1972; since then we have had further unmanned missions, and by now there is no area of the Moon which has not been explored. But let us not forget that our knowledge is based upon the work of the great selenographers of the past. All honour to them.

8

Features of the Moon

Look at the Moon through a telescope, and the first impression is one of total chaos. There is so much detail that the newcomer to observation will feel bewildered. Yet before long the impression wears off, and order begins to emerge. A few evenings at the telescope – or, for that matter, studying photographs – is enough for the main features to be identified. Mountains, peaks, craters, valleys and lowlands all have their own special characteristics, and no two are exactly alike.

It is easy to be deceived by the rapid changes in appearance due to the shifting angle of illumination. A peak or a crater can alter beyond recognition in only a few hours, simply because we depend so much on shadow. When the Sun is rising or setting over an elevation, the shadow cast is long and prominent, but under a high Sun the shadow shortens, or even disappears altogether, just as the shadow of a post will do as the Sun rises over it. As there is no local colour on the Moon, the feature will be almost unidentifiable unless it is markedly darker or brighter than its surroundings. The effect is most striking with the craters. In fact, full moon is the very worst time to begin observing. I had my first telescopic view when the Moon was very near full, and I was frankly bewildered (mind you, I was only seven years old at the time, so perhaps I can be forgiven).

A walled formation is at its most prominent when it is on or near what is called the terminator, and its floor is shadow-filled. The terminator is the name given to the boundary between the day and night sides of the Moon. It should not be confused with the limb, which is the Moon's apparent edge as seen from Earth. The limb remains in almost the same position, though it does shift slightly because of the various librations. On the other hand, the

| Crescent | Half | Gibbous | Full |

Fig. 23. The Moon's phases

terminator sweeps right across the disk twice in each lunation, first when the Moon is waxing ('morning terminator') and then when it is waning ('evening terminator'), so that even an hour's watch will show up definite movement. In fig. 23, the full, gibbous, half and crescent phases are given, with the limb drawn as a continuous line and the terminator dotted.

Owing to the roughness of the Moon's surface, the terminator does not appear as a smooth line. As the Sun rises, the first rays naturally catch the mountain-tops and high areas before the valleys and crater-floors, so that the terminator presents a very jagged and uneven appearance. Peaks glitter like stars out of the blackness while their bases are still shrouded in night, so that the summits appear completely detached from the main body of the Moon; ridges make their first appearance in the guise of luminous threads, while a crater will show its rampart-crests and the top of its central mountain while its floor is still perfectly black. On the other hand, a low-lying area will look like a great dent in the terminator, and take on a false importance for a few hours. Even with a small telescope it is fascinating to watch the slow, steady progress of sunrise over the bleak lunar landscape.

The result of all this is that the features shown on lunar maps cannot be seen properly all at the same time. In fact, it is true to say that full moon is the very worst time for the beginner to start observing, because the limb appears smooth and complete all round the disk, and the shadows are at their minimum. Also there are the bright rays, which drown most of the other details – so that the Moon becomes a speckled, confused circle of light.

I am an old-fashioned observer of the Moon (indeed, I was once referred to as a lunar dinosaur), and the present chapter is given over to old-fashioned observation; after all, when I began looking at the Moon the whole concept of space-flight was equated with astrology, alchemy, perpetual motion and schemes for extracting large quantities of gold from sea-water. As we have noted, all astronomical telescopes give inverted images, so that to a northern-hemisphere observer the Moon's south pole will be at the top. For this reason I have given the maps in this book with telescopic orientation. Anyone who lives in the southern hemisphere, or who wants to be thoroughly modern, need do no more than turn the pictures upside-down.

The question of 'east' and 'west' is different. Before the Space Age, when all maps were 'south-up', it was conventional to say that east was on the right, west on the left, so that the well-formed Mare Crisium was near the west limb and the dark-floored crater Grimaldi was to the east. Rocket men wanted everything to be oriented 'north-up', and to reverse east and west, thereby confusing folk such as myself. Finally there was a debate at the General Assembly of the International Astronomical Union, ending in a vote. I spoke

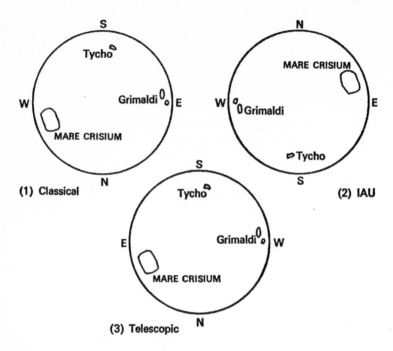

Fig. 24. Various systems of orientation. (1) Classical: south at the top, Mare Crisium to the west. (2) IAU: Mare Crisium to the east. (3) The telescopic view as used in this book. I have retained south at the top, but have accepted the IAU ruling of putting Mare Crisium *east* and Grimaldi *west*

on behalf of the defence, but I was badly defeated, and we must of course bow to the IAU ruling: Mare Crisium to the *east*, Grimaldi to the *west*. I hope that the diagram (fig. 24) will make this clear.

Next there is the question of nomenclature. Some years ago the situation became rather out of hand, as I mentioned earlier. Riccioli's system of using the names of famous men and women was always followed, but other Moon-mappers added names of their own, and at one time there were several systems in operation. The IAU has now standardized matters, and the official list is followed by everyone.*

Most small-scale observers' maps of the Moon are drawn to mean libration, and this is the method I have adopted here. Actually, libration causes noticeable alterations in aspect. Sometimes the Mare Crisium seems to be very close to the Moon's limb, while at other times it is well on to the disk. Areas in the libration zones are notoriously hard to map, and it is only too easy to

---

* The whole matter caused an amazing amount of controversy, some of which, for reasons I never understood, was acrimonious. Even astronomers are human, with human weaknesses!

confuse a crater with a ridge. Not until the Space Age did we have really reliable charts of these foreshortened regions – and of course the far side of the Moon, which does not concern us in the present chapter.

The result of all this is that the features shown on the maps cannot be seen properly all at the same time. Moreover there are the bright rays, which dominate the entire scene near full phase and drown all other details.

Of course the vast dark plains known as seas catch the eye at once, whatever the state of illumination. They take up much of the visible disk, and cover a large part of the western hemisphere – which is why the last-quarter moon, when the western side is shining on its own, is much less brilliant than the first-quarter moon, when the eastern half is visible.

On the Earth-turned side of the Moon there are nine important seas and a number of lesser ones, though nearly all of these are combined into one great connected system in the same way as our water-oceans, and usually there are no hard and fast boundaries. The names are romantic indeed. Latin is still regarded as the universal scientific language; thus the Sea of Showers becomes 'Mare Imbrium', the Sea of Vapours 'Mare Vaporum' and so on. (*Mare*, I need hardly add, is Latin for sea; plural *maria*.) We also have a Sea of Nectar (Mare Nectaris), a Bay of Rainbows (Sinus Iridum), an Ocean of Storms (Oceanus Procellarum) and so on – but we have to admit that the names are distinctly inappropriate. Showers, rainbows, nectar and storms are unknown on the Moon.

The seas, with their Latin names and English equivalents, are listed in Appendix 6. The Latin versions are used by astronomers all over the world, and I propose to keep to them here. It is not as though they are in any way difficult; one soon becomes used to them.

We know, of course, that there is no water on the Moon now, and the so-called seas are dry plains without a trace of moisture in them. Analyses of the rocks brought home by the astronauts seem to show that there has never been much water on the Moon; the maria were once lakes of liquid lava, and this has solidified to make the volcanic rock known as basalt. Undoubtedly the maria were viscous well after the adjacent regions had become permanently rigid. There is obvious proof of this in the features bordering them. We can see traces of the old boundary between the Mare Humorum and the Mare Nubium (Third Quadrant); the mountainous border between the Mare Imbrium and the Mare Serenitatis has been widely breached, while various 'coastal' craters, such as Le Monnier and Doppelmayer, have had their seaward walls ruined and levelled. Lava is a great destroyer.

Many of the major seas are more or less circular, with ramparts that are in places lofty and mountainous. Look, for instance, at the most impressive of all – the Mare Imbrium (fig. 25, see p. 73), in the Second Quadrant. It seems oval in shape, but this is because it is foreshortened; really it is almost circular, hemmed in by four mountain arcs, and large enough to hold Britain and France put together.

The foreshortening effect is even more striking with the Mare Crisium, not far from the eastern limb. This is a conspicuous sea, visible with the naked eye and separate from the main system. It measures 260 miles in one direction and 335 in the other – but the longer diameter is along the east–west axis, which comes as a surprise to the unwary observer. Orbiter and Apollo photographs of it taken 'from above' show it in its true form, and it is distinctive and well defined; but nobody, apart from the Apollo astronauts, has yet had a direct view of it as it really is. The third of the almost circular, well-bordered seas is the Mare Serenitatis, almost exactly equal in area to Great Britain; it lies in the First Quadrant, east of Imbrium. Less perfect, but still basically circular, are the Mare Nectaris and the Mare Humorum, which are smaller structures on the borders of major seas.

The only 'sea' which extends well on to the far side of the Moon is the Mare Orientale. I have a fatherly interest in it, because I actually discovered it – long before the Space Age – when I was mapping the libration zones with the modest telescope at my own observatory, then at East Grinstead in Sussex. I also proposed its name; since it was close to what was then regarded as the east limb of the Moon, I suggested Mare Orientale, the Eastern Sea. (Thanks to the IAU decision, it is now on the *western* limb!) When I came across it I had no idea what it really was; I took it for a minor mare of the Marginis type, because only a small fraction of it can ever be seen from Earth. In fact it is a vast ringed formation with concentric mountain borders. Otherwise, the Moon's far side lacks major maria, so that there is a real difference between the two hemispheres. It may be worth noting that Grimaldi, the dark-floored crater near the western limb of the Moon, has a diameter more than one-third that of the Mare Crisium, and if it had been better placed on the disk it would probably have been classed as a minor sea.

Many of the lunar peaks are lofty, and reach up to well over 20,000 feet above the adjacent landscape, but there are several things to be borne in mind before we can prove that the Moon's mountains are higher than ours, so let us examine the position rather more closely.

Consider the Mare Imbrium, the largest of the circular-type seas. Here there are various mountain borders. The Apennines separate the Mare Imbrium from the smaller, dark-hued Mare Vaporum, and they make a magnificent spectacle when seen at their best, at around the time of half-moon: their peaks cast long, sharp shadows across the plain, and they give the impression of a true range. The highest peak, Mount Huygens, rises to almost 20,000 feet, and there are other crests of at least 15,000 feet; the whole chain is over 600 miles long, stretching from Mount Hadley in the north to the grand crater-ring of Eratosthenes in the south. (The Apollo 15 astronauts landed near Hadley, and their Lunar Rover still stands there, waiting patiently for somebody to go and collect it.)

The Apennines break off near Eratosthenes, and there is a wide gap, so that the Mare Imbrium links up with the even vaster though less perfect Oceanus Procellarum. Then comes another range, the Carpathians, much lower than the Apennines and rising to no more than 7,000 feet anywhere, but over 100 miles long. When the Carpathians come to an end, there is another broad gap in the Mare border.

In the north, the Mare Imbrium is bounded by the Alps, which do not however join up with the Apennines; there is a gap between the two ranges, so that the sea-floor is connected with that of the neighbouring Mare Serenitatis. (There is a difference in level, and also in age.) The Alps are by no means the equal of the Apennines, but they have an interest all their own, largely because of the presence of the remarkable Alpine Valley. Near here, too, is the dark-floored crater Plato, one of the most famous formations on the Moon. Further west lies the Sinus Iridum or Bay of Rainbows, bordered by the Jura Mountains; when on the terminator, so that the Juras catch the sunlight while the Bay is in darkness, the 'jewelled handle' effect is superb. Beyond the Bay, the mountain border is resumed, up to the abrupt ending near the Sinus

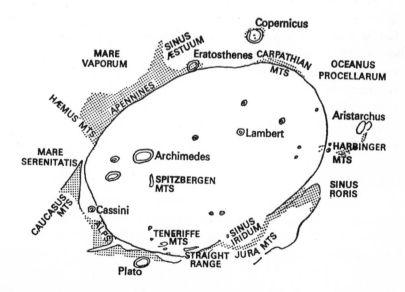

Fig. 25. The mountain walls of the Mare Imbrium

Roris or Bay of Dews. The other surviving part of the Mare border is represented by the comparatively modest Harbinger Mountains, which do not make up a definite range, but are better described as clumps of hills.

The general picture is shown in fig. 25 – and one fact stands out at once. The Apennines, Alps, Carpathians, Juras and probably the Harbingers all make up part of the circular wall of the Mare Imbrium. The Mare itself is depressed below the outer country, and it seems that we are dealing with a 'crater' on a tremendous scale. In other words, the mountain ranges are 'crater walls', and not mountains of Earth type.

The same is true of the borders of the Mare Serenitatis, where we have the Caucasus and Hæmus ranges; of the Mare Humorum, where we find the unofficially named The Percy Mountains, and so on. Mountain chains on the bright areas are lacking. Even the Altai range, which runs south-westward from the crater Piccolomini, is not properly a chain of mountains, and is now known officially as the Altai Scarp.

The chief ranges on the Moon are not of the same type as our Himalayas or Rockies, and are essentially the boundaries of the regular maria. Yet there are smaller ranges which do not make up the boundaries of seas. Of these, the most intriguing is the Straight Range, near Plato in the Mare Imbrium, a little chain 40 miles long and rising to less than 6,000 feet, but curiously regular, with an abrupt beginning and an equally abrupt end. Also notable are the Riphæan Mountains near the small, bright crater Euclides, which have altitudes of less than 400 feet, and may possibly be all that is left of the rampart of a destroyed crater.

Whatever view we take, the Moon's ranges are of the greatest significance in theoretical studies, but we must beware of drawing too close an analogy with the familiar mountain chains of the Earth.

There are countless separate peaks on the Moon. They are most numerous in the uplands, but those on the seas are the more impressive. They are not so lofty as the summits of the chain-mountains; all the same, some of them would rank with the well-known peaks of our own world. Look for instance at Pico on the Mare Imbrium, a hundred miles south of Plato – a mountain mass with broad slopes, and foothills studded with pits and craterlets. It may not seem impressive on the map, but the highest crest rises to a full 8,000 feet above the plain, so that it is twice the altitude of Scotland's Ben Nevis. Not far off is another superb mountain, Piton; and there are many other examples.

Less important peaks are very common, and the Moon is dotted with hills, some of them no more than mere mounds. Even the crater-floors are not free of them, and the smoothest parts of the maria are very far from flat, though it is naturally difficult to recognize slight differences in level unless we catch them under a low sun. Well-defined wrinkle-ridges are obvious enough when fairly

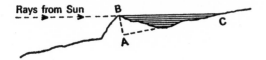

Fig. 26. Measuring the height of a lunar peak (AB) by the length of its shadow (BC)

near the terminator; there is one excellent example crossing the Mare Serenitatis. Many of these ridges seem to be the remnants of the walls of destroyed craters. And, of course, the astronauts have confirmed that the maria are rock-strewn.

On Earth the forces of erosion, such as the wearing-away of heights by wind, water and weathering in general, are constantly at work. But the Moon has no atmosphere, and from the lack of wind and water we might expect the peaks to be much rougher and more jagged. This used to be the popular view, but it has turned out to be wrong. When David Scott and James Irwin, from Apollo 15, drove across the foothills of the Apennines they had a grand view of one of the peaks, Hadley Delta. Scott called it a 'featureless mountain', and the photographs brought home show how right he was.

Galileo was the first to make a serious effort to measure lunar mountain heights. He made use of the obvious fact that the rays of the Sun will catch mountain-tops in preference to the lower-lying areas around, so that a peak may appear as a brilliant point detached from the main body of the Moon. Galileo timed how long the mountain remained lit up when on the night side of the terminator, after which its real distance from the terminator, and hence its height, could be worked out.

This is all very well in theory. Unfortunately the terminator is so irregular, because of the Moon's uneven surface, that its position cannot be measured properly, and Galileo's results were inaccurate. He believed the Apennines to be about 30,000 feet high; really they are much less than this. A better method is to measure the shadow cast by the peak itself, as shown in fig. 26. The position of the peak is known, and so is the angle at which the Sun's rays strike it at any particular moment, so that its height relative to the adjacent surface can be calculated. Of course there are various complications to be taken into account, but the method itself is straightforward enough.

Since there are no oceans on the Moon, we cannot relate the mountain heights to 'mean sea level', but at least we can be sure that the lunar peaks are higher relatively than those of the Earth.

Wherever there are mountains, there will be valleys, and this is so on the Moon. Some are mere passes, while others are really spectacular. Yet here again we must be wary of jumping to conclusions. The Rheita Valley, in the Fourth Quadrant, is long enough to stretch from London to Birmingham, and at first sight looks almost as though it had been scooped out by a gigantic chisel, but a closer look shows that it is not a true valley at all. It is a crater-chain from beginning to end, and there is nothing genuinely valley-like about it.

The Alpine Valley, near Plato, is quite different. Over 80 miles long, it is not a crater-chain; it slices right through the mountains, and is a superb sight in even a small telescope. It has no equal on the Moon, and I for one never tire of looking at it.

It is not easy to decide just what is a genuine valley and what is nothing more than a gap between two roughly parallel ridges; the term is a loose one. Many of the valleys, too, resemble that of Rheita in that they are really crater-chains.

Next there are the remarkable features known as domes. As their name suggests, they are surface swellings, and give the impression that they were produced by some internal force which pushed up the Moon's crust without being able to break it. They are by no means rare; they tend to occur in clusters. For example there are a number of domes near the crater Arago on the Mare Tranquillitatis (not too far from the point where Armstrong and Aldrin landed in 1969) and several inside the crater Capuanus, near the border of the Mare Humorum; another group lies close to Prinz in the Harbinger Mountains. One of my favourite domes is the splendid specimen near the little crater Milichius.

Many of the domes have been found to have summit pits, so that they are presumably produced by internal action rather than meteoritic impact. In 1959 G. P. Kuiper called them 'extinct volcanoes', and no doubt he was right.

The domes are so gentle of slope that they are not striking even under a low sun, but many can be seen with modest equipment. They also occur on the bright uplands, though the highland domes are naturally even more elusive than those on the grey background of the maria.

Every Earth geologist is interested in faults, and so is every student of the Moon. Examples are not lacking. For instance, there is the remarkable 'Straight Wall', near the edge of the Mare Nubium and not far from the interesting crater Thebit. The name is inappropriate, because the Straight Wall is not straight, and it is certainly not a wall. The surface of the plain to the west is almost a thousand feet lower than on the east, so that the so-called 'wall' is nothing more nor less than a giant fault. It begins at a clump of hills known commonly as the Stag's-Horn Mountains, and ends at a small craterlet 60 miles to the north. Before full moon it shows up as a dark line, since it is casting shadow to the west; it then vanishes, and for some days cannot be identified at all, though the Stag's-Horn peaks can usually be traced. The Wall then reappears as a bright line, with the slanting rays of the Sun shining upon its inclined face.

It used to be thought that the Straight Wall must be almost sheer, so that anyone standing west of it would be confronted with a near-vertical cliff. In fact this is not so; the slope is no more than 40 degrees. It is the most perfect structure of its kind, and is one of the show-places of the Moon. No doubt it will eventually become a major tourist attraction!

Ridges, too, abound on the Moon – for example the low, snaking wrinkle-ridges which cross the floors of the maria. Ridges further on to the seas are usually the walls of old 'ghost' craters which have been so completely destroyed that only fragments of their borders remain.

Before coming on to the craters, I must say something about the crack-like features known as clefts or rills (often called rilles; I prefer to keep to the original spelling). One of the most prominent of them lies near the centre of the disk, near the small but distinct crater Ariadæus. It can be seen with any small telescope when the lighting is suitable, and it looks remarkably like a crack in the Moon's surface, running for well over 100 miles. Close to it is a curved rill associated with the four-mile crater Hyginus. Most spectacular of all is the lovely winding valley-cleft running out of Herodotus, the companion-crater of the glittering Aristarchus near the junction of the Mare Imbrium and the Oceanus Procellarum. It starts inside Herodotus, broadens out into a formation which has been nicknamed the Cobra-Head, and winds its way across the plain. Then, too, there are whole systems of rills, such as that in the region of Triesnecker – again in the general area of Hyginus and Ariadæus; and some craters, such as Gassendi on the edge of the Mare Humorum and Alphonsus in the central chain of which Ptolemæus is the senior member, have elaborate cleft-systems on their floors.

In pre-Space Age days there were some ideas about rills which we now know to be wrong. R. B. Baldwin, in 1949, described the Triesnecker clefts as 'irregular cracks, very jagged in appearance, which seem to be bottomless'. In fact they are not more than a mile or two deep at most, whereas they are up to three miles wide. They are not jagged, and they are not true cracks; there is no doubt that they are collapse features, and there is no analogy with our river-beds. No water has ever flowed in them. Their distribution is not random; some parts of the Moon are riddled with rills, while other areas lack them completely.

Some of the so-called rills are in fact crater-chains; the Hyginus Rill is one of these. One rill has been seen from close range. From Apollo 15 David Scott and James Irwin drove almost to the edge of the vast rill in the Hadley region, near the Apennines.

So much, then, for the seas and the mountains, the domes and peaks, the faults, ridges and rills. Up to now I have only touched upon the craters of the Moon – and yet these craters are the dominating features of the lunar landscape. Without them, the Moon would look strange indeed.

The Craters of the Moon

Who has not heard of the craters of the Moon? They dominate the entire lunar scene, and no area is free from them. They are found on the plains, in the rugged uplands, on mountain-tops and on the floors and walls of larger structures. They range in size from vast enclosures well over 100 miles in diameter down to tiny pits, so small that from Earth they cannot be seen at all. As the astronauts walked (or drove) across the Moon they found crater-pits everywhere.

In fact the term 'crater' is rather misleading, since it conjures up the picture of a deep, steep-sided hole. Also, one thinks instinctively of volcanic craters such as that of the Earth's Vesuvius, but the lunar formations are not in the least like this. In shape they are more like shallow saucers than deep wells. Many of them have massive central peaks, or groups of peaks; others have floors which are relatively flat. Some have brilliant walls and bright floors, while others have dark interiors. There is endless variety.

Various schemes of classification have been proposed. For instance, the large formations with relatively low ramparts and without central peaks have been called 'walled plains', while smaller features are 'crater-rings', and so on. As this is a purely observational chapter I propose to take the easy way out and call all the walled structures 'craters', but we must always remember that they show a very wide range in both scale and form.

The scale is perhaps the most striking point about them. Consider Theophilus, one of the most impressive craters on the Moon, with high walls and a massive central mountain complex (fig. 27). It is 64 miles in diameter, and it has well-marked, terraced ramparts. Bailly, much less deep, is around 180 miles in diameter. Transfer either of these to England, and there would not be much room to spare. The South Pole-Aitken Basin, on the Moon's far side (never visible from Earth), is well over 1,500 miles across. There is nothing on Earth to match craters of this kind; our largest structures would make a very poor showing on the Moon. Meteor Crater in Arizona, for example, is less than one mile in diameter, and if it lay on the Moon it would certainly not be considered worthy of a separate name.

Appearances can be deceptive. When a crater is seen on the terminator, so that its floor is filled with shadow, it looks very deep – it could even be taken for the proverbial 'bottomless pit', but under a high sun it may become so obscure that it is difficult to identify at all. (Note, incidentally, that no central mountain equals the height of the surrounding walls.) One very prominent crater is Ptolemæus (fig. 27), which lies close to the centre of the Earth-turned hemisphere; it is hard to credit that the broken walls are nowhere more than 400 feet above the sunken floor. Ptolemæus is almost 100 miles across. As yet no astronaut has been there, but an observer standing in the middle of the floor would be unable to see the walls at all, for the excellent reason that they would be well below his horizon.

Of course, some craters are much deeper than Ptolemæus. In some cases the floors are at least 20,000 feet below the crests of the ramparts. Near the lunar poles there are deep craters whose floors are always in shadow, so that they remain bitterly cold – a point to which I will return later.

A good example of a smaller crater is Theætetus, on the Mare Imbrium. It is 32 miles across, so that its area is greater than that of the Isle of Wight, and the walls rise to 7,000 feet above the floor. This sounds quite impressive, but the floor itself is depressed 5,000 feet below the outer plain, so that anyone standing on the Mare Imbrium and looking at the rim from outside would be confronted only with the very mildest elevations. Neither are the slopes of craters steep, though it is true that astronauts have found walking uphill very exhausting – of course, space-suits are not the most comfortable of garments for long treks, even under one-sixth Earth gravity. Even with really small craters the slopes are not precipitous, so that the idea of a lunar crater as a gaping hole in the surface, banked by mountains rising almost sheer from the shadowed depths, is very wide of the mark.

There is one common factor. All craters, large or small, are basically circular, even though they may have been battered and distorted by later

Fig. 27.
Crater profiles:
Theophilus and
Ptolemæus

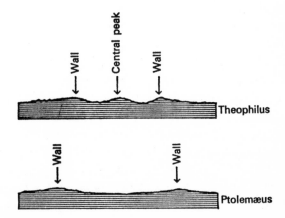

activity. As seen from Earth, the craters away from the disk appear elliptical, but this is due solely to foreshortening, and space-craft pictures give the true story. Look for instance at two photographs of the regular, dark-floored Plato, which is 60 miles in diameter and is instantly recognizable whenever it is sunlit. The Earth-based picture shows it as elliptical; the Orbiter view, taken from almost overhead, proves it to be virtually a perfect circle.

The gradation from circle to long, narrow ellipse is very evident. Consider Pythagoras near the north-west limb, Clavius in the rough southern uplands, and Arzachel in the Ptolemæus chain. All are circular – but only Arzachel looks it. Clavius is 146 miles across, so that its area is greater than that of Switzerland; a string of smaller craters crosses its floor, and the walls are some three miles above the sunken interior. It appears decidedly oval, because it is closer to the limb than to the centre of the disk, while Pythagoras is so drawn out that its interior details are hard to see at all, apart from the massive central mountain. Were Pythagoras better placed, it would indeed be imposing. It is about the same size as Ptolemæus, but much more regular and much deeper.

The largest craters – those known conventionally as walled plains – must also be the oldest. Many of them are now so broken and ruined that they are scarcely recognizable. Janssen in the Fourth Quadrant, not far from the Mare Australe or Southern Sea, must once have been a noble formation, with high continuous walls towering to thousands of feet above its sunken floor, but it has been so roughly treated by later disturbances that it is now no more than an enormous field of ruins, broken by craters, ridges, pits and rills, with its walls breached in dozens of places and completely levelled in some. Only when it is right on the terminator, and filled with shadow, does it give a slight impression of its former self.

The term 'field of ruins' has also been applied to Bailly, on the south-western limb of the Moon. It is actually the largest named crater on the Earth-turned hemisphere, but it is so badly placed and foreshortened that only the space-probe pictures can show it well. Were it darker-floored and more central on the disk, it would certainly have been classed as a minor sea rather than as an outsize walled plain.

However, quite a number of large structures have managed to escape relatively unharmed. Clavius is a case in point. Admittedly it contains several craters which are large in their own right, but the walls are fairly regular, and when the Sun is rising or setting over it the result is an apparent dent in the terminator which can be seen without any optical aid at all.

Many of the walled plains, particularly the damaged ones, are hard to identify when the Sun is high over them. The exceptions are those with dark floors, such as Plato, which Hevelius called 'the Greater Black Lake'. Grimaldi and Riccioli, near the western limb, are also extremely dark, and there are some smaller formations with similar lunabase floors; Billy and Crüger, both in the Third Quadrant, are examples.

There is a tendency for the walled plains to be arranged in lines and groups, a point to which I will return later; there is nothing random about their distribution, though certainly on the Moon's far side the situation is less clear-cut. One of the most impressive of the chains lies very close to the apparent centre of the disk. Here we have Ptolemæus, Alphonsus and Arzachel, which are magnificent when near the terminator though decidedly obscure near full moon. Ptolemæus has a reasonably darkish floor, with few conspicuous craterlets and no central mountain. Alphonsus, 80 miles across, is the middle member of the chain; there is a reduced central peak, and on the floor there is a fine system of rills, photographed in detail not only by the Orbiters and Apollos but also by the last of the Ranger probes, No. 9, which crash-landed not far from the central peak. The southern member of the trio, Arzachel, is smaller and deeper than Alphonsus, with a considerable central elevation. We have here an excellent gradation between walled plain and peaked crater, adding force to my contention that the structures are of essentially the same type and origin.

Coming now to the generally smaller formations we find that central peaks are common, though by no means the universal rule. Some craters have floor-mountains which rise to thousands of feet, though they never equal the height of the surrounding rampart; others may have lower, many-peaked central elevations, and sometimes the so-called central mountain is little more than a mound. Then, too, there are formations which have smaller craters centrally placed on their floors, and there are many examples of true concentric crater-rings. Taruntius, near the Mare Crisium, and Vitello on the edge of the Mare Humorum, are good examples.

One of the finest of all the regular craters is Theophilus, in the Fourth Quadrant near the edge of the Mare Nectaris, 64 miles in diameter and over 14,000 feet deep, with a magnificent central elevation and massive walls. It has broken into a second formation, Cyrillus, which is about the same size, but whose walls and central peak are much lower; to the south lies the third member of the chain, Catharina, where the floor is very rough and lacks a central mountain. On the Mare Imbrium we have the superb trio made up of Archimedes, Aristillus and Autolycus, where the same sort of gradation in type is shown; and there are others almost equally worthy of note. But among all the Moon's craters perhaps the most impressive is Copernicus, on the Oceanus Procellarum. It has been nicknamed 'the Monarch of the Moon', and with good reason.

Copernicus has massive walls rising in places to 17,000 feet above the inner amphitheatre. The distance right across the crater, from crest to crest, is 56 miles, but the true 'floor' is only 40 miles across, since the rest is blocked with rubble and débris produced by huge landslides from the ramparts. The central heights are made up of three distinct, multi-peaked masses, while lower hills and mounds litter the floor. There is nothing smooth about the interior of Copernicus. The walls are terraced — a common feature of lunar craters, but particularly well marked here.

In 1966 the space-probe Orbiter 2 photographed Copernicus from a mere 28 miles above the Moon's surface. The crater, approximately 150 miles away, was shown obliquely, and the result was termed 'the Picture of the Century'; it was the most magnificent lunar view obtained up to that time, though of course we have since had far more detailed images from the later space-craft, notably Clementine and Prospector. The most delicate details are not visible from Earth, even with our best telescopes, but a modest instrument will show that Copernicus is a sight never to be forgotten. Moreover, it is the focal point of a system of bright rays second in importance only to that of Tycho, though they are of rather different type.

Copernicus, with a probable age of perhaps a thousand million years, is one of the youngest of all the major craters, and has escaped damage, but other formations have not been so lucky. Those on the 'coasts' of the maria have had their seaward walls battered down and levelled, so that they have been turned into huge bays; Fracastorius on the border of the Mare Nectaris and Le Monnier on the edge of the Mare Serenitatis are good examples, and the same may be true of the famous Sinus Iridum or Bay of Rainbows. In some cases the ruins of a seaward wall can still be seen, and even, as with Hippalus on the border of the Mare Humorum, the wreck of a central mountain.

The Sinus Iridum, most splendid of all the bays, leads off the Mare Imbrium. The ground level drops gradually to the west, and low, discontinuous remnants of the old east wall can still be traced between the two jutting capes to either side of the strait which separates the Bay from the main Mare. When the terminator passes close by, the mountain peaks of the western border (the Juras) catch the light, producing the unique 'jewelled handle' effect.

Old craters right on the seas have been even more reduced, and have been drowned by the mare material, so that they now show up as ghosts — marked sometimes by low, discontinuous walls, sometimes by ridges, sometimes by nothing more than a slight change of tint on the plain.

It is tempting to go on describing crater after crater, since each has its own special points of interest. I shall have more to say about some of them later – notably Aristarchus, which is a mere 23 miles in diameter, but is the most brilliant feature on the entire surface, so that it shows up even when illuminated only by earthshine. There is plenty of variety on the Moon.

Craterlets, with diameters ranging from more than a dozen miles down to a few yards or even less, pepper the whole Moon. Some are true miniatures of larger craters, even to the central hills; others are pits, with depressed floors but virtually no walls rising above the outer level (often called 'blowhole craters'). Probe pictures, and above all the photographs taken by the astronauts on the surface itself, show that there are countless craterlets much too small to be seen from Earth.

The lining-up is even more marked with the small formations than with the larger ones. Many of the so-called rills are made up basically of small craters which have run together, often with the loss of their dividing walls. Others, such as the Hyginus Cleft, are crater-chains in part, though it is true that there are plenty of rills which show no trace of crater-like enlargements. Crater-chains are very common indeed, and there is nothing surprising in this. We have a complete series from the 'giants' down to the 'strings of beads'.

Here and there we come across real lunar freaks. Perhaps the most remarkable of all is Wargentin, close to the large walled plain Schickard in the Third Quadrant. Its floor is not sunken, but is raised above the outer surface by about a thousand feet. What may have happened is that some blockage caused the molten magma to be trapped inside the amphitheatre when the crater had just been formed, so that instead of subsiding and draining away, as usually happened, the magma solidified where it was. If so, the true floor of Wargentin is hidden, and all we can see is the top of the lava-lake. In places the floor is level with the top of the rampart, but in other areas there are still traces of a wall – one segment rises to as much as 500 feet. Still, the general impression is that of a flat-topped plateau, not a conventional crater. Wargentin is large enough to hold the whole of Lancashire, and it is a pity that it is not more central on the disk, as there are no other plateaux anything like as large. Various smaller specimens exist, but they are neither so perfect nor so prominent.

Finally there are the bright rays, which dominate the whole scene when the Moon is near full. Unlike most other details, they are best seen under a high light; they are very obscure when near the terminator, and become conspicuous only when the Sun has risen to a considerable altitude over them. Of the many ray-systems on the surface, two stand out as being incomparably more splendid than the rest: those of Tycho and Copernicus.

Tycho is a well-formed crater in the southern uplands, 54 miles across, with high terraced walls and a central mountain complex. Magnificent though it is, Tycho lies in so crowded an area that it would not be outstanding were it not for the rays. When it first emerges from the long lunar night it seems to be a perfectly normal bright crater, but gradually the rays come into view, until by full moon they dominate the whole of the southern part of the disk. There are dozens of them, streaking out in all directions from Tycho as a focal point; they cross craters, plains, peaks and valleys, uplands and maria, rills and pits without showing obvious deviation.

Near full moon, when Tycho's rays dominate the whole scene, it is tempting to believe that the crater lies at the lunar pole. In fact it does not; it is some way away.

Rather unexpectedly, the rays cannot be traced inside Tycho itself. There is a ray-free area round the rampart, showing darkish under a high light, where the streaks stop short; neither do they radiate from the exact centre of Tycho, since many of them are tangential to the walls. Yet there can be no doubt that the rays were produced at the time of the impact which formed the crater – and since the rays cross all other formations this is proof positive that Tycho must be the youngest structure in this part of the Moon. One ray stretches right beyond the Mare Serenitatis, passing close to the bright little crater Bessel, and there has been a great deal of discussion as to whether this is one long, genuine 'Tycho ray' or whether it has been renewed along its course – though it is not easy to see just how this could have happened. In January 1968 the Surveyor 7 space-craft made a gentle touch-down on the outer slopes of Tycho, and sent back excellent pictures. It is still there, and one day, no doubt, it will be collected and taken away to a lunar museum.

The rays associated with Copernicus are different from those of Tycho. They are not so luminous, and at full moon, when they are at their best, they appear less brilliant than the gleaming crater-ring of Copernicus itself. Neither are they so long or so regular as those of the Tycho system, though they spread widely over the surrounding plain.

Here and there over the disk other ray-centres can be made out: Kepler on the Oceanus Procellarum, Olbers close to Grimaldi in the far west, Anaxagoras in the north, and so on. Some craters have rays which are so dark as to be scarcely detectable. We also find small craters surrounded by bright patches – Euclides, near the Riphæan Mountains, is an example – and craterlets with short ray-systems. Near full, the various rays confuse the whole lunar scene so thoroughly that even the practised observer may have trouble in finding his way about.

There are definite differences between the Earth-turned and far hemispheres of the Moon. For example, on the far side there are large, light-floored enclosures known as palimpsests, which are not evident on the familiar hemisphere – and of course there is a notable lack of maria; of the main 'seas' only the Maria Orientale extends on to the far side.

This, then, is the lunar picture – a scene of apparent chaos which, on close examination, turns out to be not quite so chaotic after all, and where craters and peaks mingle with valleys, faults, domes, rills and mountain-chains. Lunar landscapes are indeed fascinating, and anyone equipped with a small telescope can spend many hours enjoying them.

10

The Past and Future Moon

Craters are almost always associated with volcanoes: Vesuvius, Etna, Stromboli and many others. Some are violently explosive, while others are gentler. Go to Halemaumau, atop Mauna Loa in Hawaii, and it is (usually) quite safe to look down into the lava-lake; believe me, it is a fascinating sight. Obviously, then, it seemed logical to assume that the lunar craters were of volcanic origin, similar to our calderæ, and for many years most people believed this. I certainly did, for the reasons given in the earlier editions of this book. Sadly, I now have to accept that I was wrong. There has been extensive vulcanism on the Moon, but the craters were produced by meteorites which rained down on the Moon thousands of millions of years ago, during the period of what has become known as the Great Bombardment.

Before going into this, let us pause briefly to say something about past theories – of which there are many, some outwardly convincing and others frankly weird. As a start, there were the various 'ice theories'.

During the nineteenth century it was widely believed that the Moon was a frozen world, where the surface temperature was always very low. One man who disagreed was the fourth Earl of Rosse, who constructed a sensitive piece of equipment and used it to show that during the long lunar day the temperatures of the equatorial regions, at least, are very high indeed.* But Lord Rosse's results were not generally accepted until well after his death, in 1909, and there was strong support for a 'glaciation' theory, originally proposed by a Norwegian named Ericson in 1885 and supported four years later by S. E. Peal, a tea-planter and amateur astronomer, who wrote an entire book about it. According to Peal, the lunar maria are frozen water-lakes, and the whole Moon is coated with a layer of ice. Water vapour sent out from below the crust produces a local, dome-shaped atmosphere, and this freezing material condenses around the vents, forming icy craters. Next in line came the Austrian engineer Hans Hörbiger, who took up the idea in 1913 with the

---

* The third Earl of Rosse built a remarkable 72-inch reflecting telescope at Birr Castle, in Ireland, in 1845; with it he obtained the first detailed views of the 'spiral nebulæ', now known to be external galaxies similar to our own. His son, the fourth Earl, concentrated upon measuring the tiny amount of heat sent to us by the Moon. After his death the telescope fell into disuse, but I am glad to say that it has now been restored, and is in full working order again.

maximum possible energy. Hörbiger's book *Glazial Kosmogonie* is one of the greatest classics of crank science, rivalled in modern times only by Immanuel Velikovsky's *Worlds in Collision* and its successors. It was Hörbiger who produced WEL (Welt Eis Lehre, or Cosmic Ice Theory) which became a powerful cult when Germany was ruled by the Nazis.

According to Hörbiger, everything apart from the Earth is made up of ice. The Moon's glacial covering is 150 miles deep, and the same is true of the planets (the canals of Mars are simply cracks in the upper ice-sheet). As the planets move round the Sun they are braked by the low-density hydrogen medium, and the same is true of the Moon, which is spiralling slowly downward and will hit us in the foreseeable future – long before the Earth itself plunges into the Sun and is snuffed out, producing a large sunspot. Also, the Moon is not the first of our satellites. There have been at least six previous moons, all of which have eventually collided with us, causing cataclysms which may be linked with past events such as the abrupt disappearance of the dinosaurs, 70 million years ago. Our present Moon was captured by the Earth between 13,000 and 14,000 years ago, and led to earthquakes and eruptions, one of which submerged the island of Atlantis. Another episode linked with the theory is, needless to say, the Biblical Flood.

How anyone could take this rigmarole seriously is hard to understand, but plenty of people did so, and during the 1930s the German Government had to issue a statement to the effect that it was still possible to be a good Nazi without believing in WEL. Moreover, there was one very serious lunar observer who became a Hörbiger devotee. This was Philipp Fauth, who published an elaborate map of the Moon and wrote several books on the subject. Fauth, like Peal so long before, believed the craters to be lakes of frozen water, and he agreed with Peal's words: 'As the lakes slowly solidified in the cooling crust, the water vapour rising from them formed a local, dome-shaped atmosphere, which became a vast condensed snowy margin and piled as a vast ring.' The maria, then, were actual sea-surfaces which had solidified.

Fauth, unlike Peal, thought that the ice had come from space in a sort of cosmic storm. He discounted the measurements which show that at noon the temperature at the lunar equator can exceed 200 degrees F, which does not seem very suitable for the permanent existence of either ice or snow; he also disregarded the obvious fact that an icy crater-rampart would not keep its shape for long, and would flatten out under its own weight. Fauth died in 1943, and so far as serious science was concerned WEL died with him.

The ice theory may seem peculiar, but an even more remarkable idea was proposed in 1942 by a certain Herr Weisberger, of Vienna, who solved the whole problem very neatly by denying that there are any mountains or craters at all. He attributed the surface markings to storms and cyclones in a dense lunar atmosphere, and was most offended when the astronomical world failed to treat him with due respect.

It was of course difficult to surpass Herr Weisberger, but a Spanish engineer, Sixto Ocampo, did so in 1949, when he announced the Atomic Bomb theory. After explaining that the Moon does not rotate on its axis (!), Señor Ocampo went on to prove that the Moon used to be inhabited by technologically-advanced beings who indulged in an orgy of nuclear war, and destroyed their civilization, producing huge craters in the process. He added that the craters were of different types because the opposing sides used different types of weapons, one species of bomb producing a crater with a central peak and another kind a crater with a flat floor. The Alpine Valley and the Straight Wall were engineering works. After the fall of the bomb which produced the ray-crater Tycho, the lunar seas were 'fired' and expelled at great speed, falling back on the Earth and causing Noah's Flood.

Señor Ocampo presented his paper to the Academy of Arts and Sciences at Barcelona, and was most annoyed when they declined to publish it. It was eventually printed in a small South American periodical, together with a letter in which the author complained that an unscrupulous British writer had stolen his theory and was planning to publish it as original work, thereby depriving Spain of the glory of the discovery. Ocampo's life-work was thereby brought to a successful conclusion, and he died almost immediately afterwards.

It should not be thought that Herr Weisberger and Señor Ocampo have had the monopoly of weird ideas. Remember, there is still an International Flat Earth Society, whose members maintain that both the Earth and the Moon are shaped like gramophone records, while the members of the German Society for Geocosmical Research believe that the Earth is the interior of a hollow globe, so that the Sun is inside it. (I once put these two societies in touch. The resulting correspondence was most enlightening, but hardly relevant here.) Space forbids any discussion of the theories of D. P. Beard, who in 1917 proposed that the craters are coral atolls, or of Dr Immanuel Velikovsky, still something of a cult-figure in America, who believed the planet Venus to be an ancient comet which bounced about the Solar System in past ages, periodically bypassing the Earth and the Moon, and causing phenomena such as (of course) the Biblical Flood. So let us turn to ideas which, although completely wrong, are not absolutely outlandish.

There are various tidal theories, of which the best known was outlined by the Bulgarian astronomer, N. Boneff, in 1936. According to Boneff, the craters were formed when the Moon's crust had just solidified, and the Moon, much closer to the Earth than it is now, was rotating on its axis comparatively quickly. The hot, viscid interior was much more affected by the tidal pull than the thin crust, and so at each rotation of the Moon on its axis the molten lava surged upward, breaking through the weak points in the crust. The action was rather like that of a pump, so that gradually the large craters were built up. As the Moon receded and the axial spin slowed down, the tidal effects lessened, and the formations produced were smaller. At last the crust became too solid to be broken by the surging magma inside, so that crater-building ceased altogether.

Boneff explained that the Earth's crust was not then solid enough to register any similar craters, though he did not rule out the possibility that the Moon still influences the numbers of earthquake shocks. Moreover, if it is agreed that the Moon will one day approach the Earth once more, Boneff maintains that it may yet be capable of covering our lands and drying sea-beds with craters before it is itself torn apart by the Earth's pull. The last paragraph of his paper reads: 'An Earth without a Moon, surrounded by a ring of minute bodies and entirely covered with formations of the lunar type, except perhaps at the poles – that is the probable state of the Earth–Moon system, if it still exists, after many thousands of millions of years.' It is a sombre picture, but it cannot be taken seriously. Quite apart from anything else, the time-scale is wrong.

Having 'cleared the air', so to speak, I turn now to the various volcanic theories, according to which the lunar craters are basically of the same nature as our own calderæ.

The first serious attempt at a comprehensive volcanic theory was made in 1874 by two English amateurs, James Nasmyth and James Carpenter. (Nasmyth's telescope is now on exhibition at the Science Museum in South Kensington. He was also the inventor of the steam-hammer.) In their book, which has become something of a classic, Nasmyth and Carpenter pictured a central volcano erupting violently and showering débris round it on all sides, so that the matter ejected from the central orifice built up the crater wall (fig. 28). As the eruptions became less violent, inner terraces were

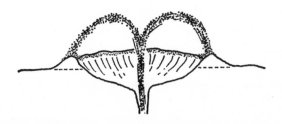

Fig. 28. Nasmyth and Carpenter's Volcanic Fountain

formed, and in the dying stages of activity, when the explosions were only just powerful enough to lift the material out of its vent, the central peak was built up. Craters without central peaks were produced when the explosions ceased suddenly, so that the floor was covered with lava which welled up from inside the Moon.

The idea looks intriguing at first sight. The terraces, the hill-top craters, the flooded plains and even the famous plateau Wargentin are accounted for, and the ringed formations do give the superficial impression of having been built up in this way. Unfortunately, there are any number of fatal objections. It is beyond all belief that a circular wall over 100 miles in diameter, and sharply defined (as with Clavius, for instance), could have been formed in this way; the slope-angles are wrong, and the walls are far too massive. Moreover, in a lunar crater the central peak is always considerably lower than the rampart, which would not be expected on the fiery-fountain theory. The explanation given for the bright rays is equally untenable. Nasmyth and Carpenter thought that the crust of the Moon had cracked in places, much as a glass globe does when it is struck, and that lava had oozed out of the cracks, forming the rays. This may have sounded convincing in 1874, but it certainly does not in 2000! Yet although Nasmyth and Carpenter were so completely wrong, they rendered lunar science a service by publishing their book at a time when interest in the Moon was at a comparatively low ebb.

Much earlier – in 1665 – an entirely different process had been suggested by no less a person than Robert Hooke, contemporary and rival of Newton. Hooke made some drawings of lunar craters which were remarkably good considering the low-power telescopes which he had to use, and he believed that the craters might be the remains of great bubbles, which rose during the period when the lunar surface was molten, bursting and leaving a solidified rim behind. Actually, a bubble of the size needed to form a really large crater can be discounted on dynamical grounds, and it is only too easy to be misled by a superficial resemblance.* But though bubbles are out of the question, uplift and subsidence processes are not, and this brings me on to the caldera analogy.

A caldera is a volcanic crater caused by the collapse of the surface into an underground cavity, either by the rapid eruption of large quantities of magma or else by the withdrawal of magma from a chamber below. There is a caldera on the top of Vesuvius, known as Monte Somma, and there are others in Africa, associated with the Great Rift Valley in Kenya. They are also found on

---

* At a very serious NASA conference in the United States, I was once guilty of showing a slide of a 'lunar scene' which was, in fact, a photograph of the top of a layer of boiling coffee. I understand that at least some members of my audience were momentarily puzzled.

Mars. The highest Martian volcano, Olympus Mons – three times the height of Everest – is crowned by a caldera 40 miles across.

The idea that the lunar craters (and the regular maria) were caldera-like structures was developed in the mid-twentieth century by a well-known American geologist, J. E. Spurr, and met with wide support. I was a firm believer in it, and only during the past few years have I been forced into the rather reluctant conclusion that I was wrong. Volcanic structures on the Moon are small and rare; there is no longer any serious doubt that the main craters were formed by meteoritic impact. But before going further, it seems worth pausing to give you the picture as it was described by Spurr.

Large calderæ were formed at a very early stage in the Moon's history, producing basins which have now been obliterated in their original form even though the depressions were later covered by lava-flows to produce irregular seas such as the Oceanus Procellarum. Then came the formation of the present-day regular maria, beginning with the Mare Nectaris and similar basins, ending much later with the creation of Mare Imbrium and, last of all, the Mare Orientale. Crater-building began by the same sort of process, and this explains the fact that there are chains of formations aligned with the Moon's central meridian as seen from Earth – because the axial rotation was already synchronous, and lines of weakness in the lunar crust were influenced by the gravitational pull of the Earth.

Many of the early walled plains were simply miniature maria; for example the dark-floored Grimaldi, near the western limb, would have been ranked as a small sea if it had been better placed for observation. With these lava-flooded craters, any central peaks would have been destroyed by the welling-up of lava, though later craters, formed when the crust had become thicker, retained central peaks and unflooded floors. There is no problem in explaining the chains, the occasional plateaux of which Wargentin is the best example, or the twins and groups; we can also account for the bays such as Hippalus and Doppelmayer on the Mare Humorum and Le Monnier on the edge of the Mare Serenitatis, where the seaward walls have been so reduced in height that they have become discontinuous. The mountain border between the Mare Imbrium and the older Mare Serenitatis was badly damaged, but was so massive that it managed to survive apart from one stretch between the modern Apennines and Caucasus.

When one crater breaks into another it is the rule that it is the smaller crater which is the intruder. Look for example at Thebit near the edge of the Mare Nubium (fig. 29). Its wall is broken by a smaller crater, Thebit A, which is itself broken by a yet smaller structure, Thebit F. This is to be expected on the volcanic theory; the oldest eruptions would be the most violent. The rule

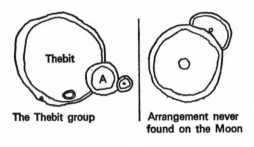

The Thebit group

Arrangement never found on the Moon

Fig. 29. Crater arrangement: (*left*) small craters intruding into larger; (*right*) a large crater intruding into a smaller peaked crater in a way that is unknown on the Moon

holds good in well over 99 per cent of cases. Ray-craters, notably Tycho and Copernicus, were born at a late stage, by which time the crust had become so firm that the magma was mainly trapped below. Gradually the volcanic activity died down, and at last, perhaps a thousand million years ago, ceased altogether.

All this sounds reasonable enough. Only now, with the development of space research, have we had to abandon the caldera hypothesis. So let us turn to the impact theory, and the Great Bombardment.

The idea that the craters are due to falling meteorites was originally put forward in 1824 by Franz von Paula Gruithuisen, an energetic if rather unorthodox German astronomer. It was then more or less forgotten, revived briefly by R. A. Proctor in England during the 1870s, and then discarded again, even by Proctor himself. In 1892 came a famous paper by G. K. Gilbert, one of the leading geologists of the time, which went into considerable detail. The impact theory was finally put into its modern form by R. B. Baldwin, of the United States, in his 1949 book *The Face of the Moon*.

It has been suggested that a meteorite would cause a circular crater only if it fell straight down, and that a missile landing at an angle would produce an elliptical scar. In fact this is not true. On impact the meteorite will penetrate the Moon's crust and act in the manner of a powerful explosive, so that a circular crater will result. There are, of course, many impact craters on Earth, and though they are very puny by lunar standards we can learn a great deal from them. The best known, as already mentioned, is the Arizona Meteor Crater, which is visible from Highway 66, the main road running between the towns of Flagstaff and Winslow. It is an impressive structure, and is well worth visiting; the diameter is 4,150 feet, and the maximum depth 700 feet. The wall rises 150 feet above the surrounding desert. Svante Arrhenius, a famous last-century Swedish scientist, called it 'the most interesting place on earth' – though Gilbert, rather surprisingly, regarded it as 'a steam explosion of volcanic origin'. It was formed around 50,000 years ago, when the region was completely uninhabited. Another very well-formed impact crater is Wolf Creek, in Australia, and there are others scattered over the globe. It is widely believed that around 65 million years ago a large meteorite fell at Chicxulub in the Yucatán region of Mexico, and threw up so much débris that the Sun was blotted out and the entire climate changed – with disastrous results for the dinosaurs, which could not cope with the new conditions, and died out.

On Earth, ancient impact craters are eroded away; this is not the case on the Moon, where the structures remain visible. If we could go back in time for, say, a 1,000 million years we would see the Moon very much as it is today. Little has happened there since those remote times. Even the 'youngest' lunar craters date back to the era long before advanced life-forms appeared on Earth.

As we have noted, the Moon's age is about the same as that of the Earth, and is of the order of 4,600 million years. However, it came into existence as a separate body, the heat generated during its formation led to the melting of the outer layers, and for a time there must have been a magma ocean several miles deep. Heavier material sank, while less dense materials separated out on to the surface and, in the fullness of time, produced a crust – thicker on the far side of the Moon than on the Earth-turned hemisphere; by cosmical standards, it did not take long for the orbital period and the axial rotation period to become equal.

At that stage there were many pieces of 'débris' moving round the Sun, and the newly-formed planets swept them up. Between 4,400 million and 3,900 million years ago the Great Bombardment went on unceasingly. Look at the quiet Moon of today and it is not easy to visualize the scene at the peak of the bombardment: meteorites rained down to produce the first major basins such as those of the Mare Tranquillitatis and the Mare Fœcunditatis. Then, between 3,900 and 3,800 million years ago, came the greatest impact of all, producing the vast Imbrian basin and having profound effects all over the Moon. The crust was ruptured time and time again, and magma seeped out, flooding the basins to produce structures such as Plato and Grimaldi. The last really violent impact, producing the Mare Orientale, dates back around 3,100 million years. The lava-flows ended rather suddenly; as the outpouring slackened many craters were left relatively undamaged, and the youngest of them, such as Tycho and Copernicus, are unflooded, so that their central mountain groups are clear-cut. On the far hemisphere, with its thicker crust, there was less flooding, which explains the absence of major maria and the presence of the light-floored palimpsests. One interesting far-side crater has been named in honour of Tsiolkovskii, the rocket pioneer who was writing about space-travel more than a hundred years ago. Here we have a feature which seems to be a cross between a mare and a crater; it has a dark, flooded floor, but the walls are high, and there is a central peak. It adjoins a structure of about the same size, Fermi, which is unflooded.

It has been suggested that some basins were formed even before the onset of the Great Bombardment, and that these were filled during the lava-flow period, accounting for maria such as Frigoris and Australe, which are of no particular shape. Small volcanic features such as the Hyginus Rill – really a chain of craterlets – were latecomers. And, as we have already seen, the major ray-centres, such as Copernicus and Tycho, came near the end of the Moon's active life.

There is no reliable report of a crater having been formed during the last few centuries, when the Moon has been under observation. One claim, dating back to 18 July 1178, can be dismissed out of hand. According to a British monk, Gervase of Canterbury, the crescent moon was seen 'to split in two... a flaming torch sprang up, spewing out over a considerable distance fire, hot coals and sparks. Meanwhile, the body of the Moon which was below writhed... and throbbed like a wounded snake.' Fairly obviously this indicates an ordinary cloud phenomenon, even assuming that there is any truth in the report. Yet it has been seriously suggested that this curious disturbance was due to the formation of an impact crater on the Moon's far side – and the feature itself was taken to be the 14-mile crater which has been named Giordano Bruno. Luckily the good Gervase gave the time of his observation – and it is found that the Moon was only 5 degrees above the horizon. I fear we have no choice but to dismiss the whole report as simply a 'Canterbury tale'.

Meteorites must land on the Moon sometimes, and no doubt produce craterlets, just as happens on the Earth, but the chances of an impactor large enough to produce a major crater seem to be very low, and true volcanic activity died away long ago. The Moon today is a calm, placid world, though at one stage in its history it must have qualified as one of the wildest places in the whole of the Solar System.

11

The Lunar Atmosphere

'The Moon is an airless world.' This statement is to be found in almost every book or article dealing with our satellite, and it is more or less true. There is a trace of atmosphere, but the density is so low that it corresponds to what we normally call a good laboratory vacuum. The total weight of the entire lunar atmosphere is no more than 30 tons.

There is no mystery about this. The Moon has only $\frac{1}{81}$ the mass of the Earth, and its gravitational pull is much weaker. Go to the Moon, as the astronauts have done, and you will have only $\frac{1}{6}$ of your Earth weight. Also, the escape velocity is a mere $1\frac{1}{2}$ miles per second, and the Moon has been unable to hold on to any atmosphere it may once have had.

Throw an object upward, and it will rise to a certain height and then fall back. Throw it harder, and it will rise higher. Throw it up at a speed of 7 miles per second (which admittedly would need a great deal of muscle power!) and it would never come back at all; the Earth would be unable to draw it back, and the object would escape into space. Luckily for us, our main atmosphere is made up of particles which cannot work up to this speed, but the Moon's relatively feeble pull is inadequate, and any dense atmosphere would have leaked away quite quickly. Compare it with other members of the Solar System. Jupiter (escape velocity 37 miles per second) has been able to hold on to all its gases, even the lightest of all, hydrogen; Mars (3.1 miles per second) has only a thin atmosphere, made up chiefly of the heavy, sluggish gas carbon dioxide; Mercury (2.3 miles per second) has almost no atmosphere at all.

Less than two centuries ago astronomers such as William Herschel believed that there might be an appreciable lunar atmosphere, but there is an easy way to show that the density must be less than ours. All we have to do is to observe some lunar occultations.

As the Moon moves across the sky it often passes in front of stars, hiding or occulting them. There are a few bright stars which lie in the Zodiacal band, and can therefore be occulted; Antares in the Scorpion, Aldebaran in the Bull and Spica in the Virgin are all occulted now and then. An occultation is virtually instantaneous, because a star is to all intents and purposes a point source of light. Before occultation, the star is seen shining steadily; its disappearance behind the Moon's limb is as sudden as the flicking-out of a candle-flame in a gale. One moment the star is there; the next, it is not. Emersion at the opposite limb of the Moon is equally sudden.

Now consider what would happen if the Moon were surrounded by a dense atmosphere. For some moments before being hidden, the star's light would be coming to us after having passed through the lunar atmosphere and the star would flicker and fade. This does indeed happen when a star is occulted by the planet Venus, which is surrounded by dense and extensive atmosphere.*

An occultation of a star by the Moon is interesting, particularly if the Moon is waxing – since the occultation then takes place at the dark limb, which cannot be seen unless lit up by earthshine. When the Moon passes through a star-cluster such as the Pleiades, half a dozen naked-eye stars may be occulted over a period of an hour or two.

If a layer of atmosphere lay around the Moon's limb, it would be expected to bend or refract the light-rays coming from the star just before immersion (you can demonstrate refraction simply by shining a torch through a tank of water; the beam is obviously bent). The effect would be to keep the star in view for slightly longer than would otherwise be the case, so that the occultation would take place later than predicted. Reappearance at the opposite limb would be slightly early, so that the whole occultation would last for a shorter time than predicted by theory. The amount of refraction should give a key to the density of the lunar atmosphere responsible for it.**

This sounded plausible, and it only remained to measure the duration of the occultation. For reasons which we now know, this proved to be impossible. Sir George Airy, who was the Astronomer Royal between 1835 and 1881, believed that the effect was great enough to be detected, but later astronomers did not agree, and in fact the lunar atmosphere is so tenuous that it cannot produce any measurable effects of this kind.

When a planet is occulted by the Moon the situation is different, because a planet shows a disk, and both disappearance and reappearance are gradual. The American astronomer W. H. Pickering watched an occultation of Jupiter in 1892 and recorded a dark band crossing the planet's disk, tilted at an angle

---

* I remember seeing this well in 1959, when Venus occulted the bright star Regulus, in Leo (the Lion). Using a 12-inch reflector from Selsey, in Sussex, I saw that Venus faded and flickered very clearly well before immersion, and this made it possible to estimate the height and density of Venus' atmosphere. Results obtained later from the space-probes showed that these estimates were very near the truth.

** Refraction makes it possible for us to see the Sun and Moon before they really rise because they are 'lifted up' from below the horizon. Moreover, both the Sun and the Moon look flattened when very low down, because the bottom part of the disk is refracted more strongly than the upper part.

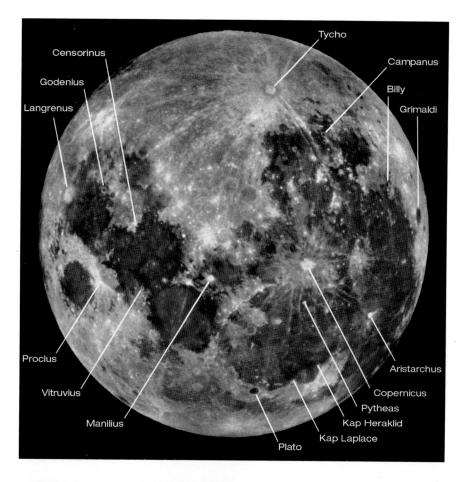

Tycho
Censorinus
Campanus
Godenius
Billy
Langrenus
Grimaldi
Proclus
Vitruvius
Aristarchus
Manilius
Copernicus
Pytheas
Kap Heraklid
Plato
Kap Laplace

Above: Photograph of the full Moon showing some of the important features. Note the brilliant ray crater Tycho and the dark-floored Plato. To the extreme right is Grimaldi, with its very dark floor.

Left: Photograph of the Earthshine, taken by Commander Henry Hatfield. Features on the Earthlit portion are easy to see.

Right: The planet Saturn
about to be occulted by
the Moon. Note how small
Saturn appears compared
with the Moon. Photograph
by Commander
Henry Hatfield.

Below: Photograph of the
Mare Imbrium, bounded in
part by the Apennines.
The three craters above
the centre of the picture
are Archimedes,
Aristillus and Autolycus.
The dark-floored crater,
Plato, 60 miles across, is at
the bottom of the picture.

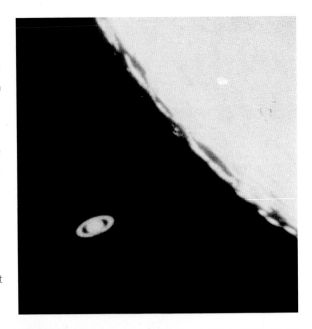

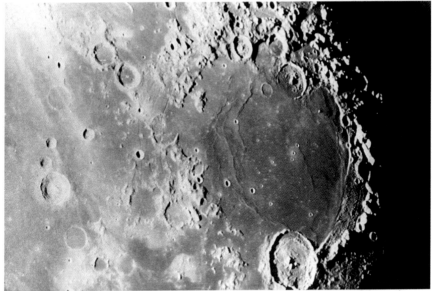

Top: Part of the Mare Nubium. To the upper right is the flooded crater Pitatus.

Above: Photograph of the Mare Humorum, with the crater Gassendi at the bottom. Photograph by Patrick Moore.

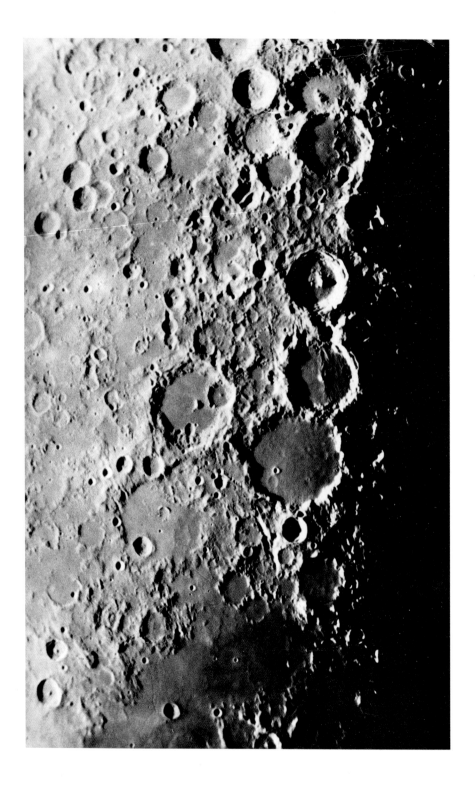

Above: The great walled plains. Near the centre of the picture is Petavius, above and to the left is Furnerius and near the bottom right is Langrenus.

Right: The great ray crater Copernicus. Above is the crater Reinhold, with its smaller companion. The mountains below Copernicus are the Carpathians.

Opposite: In this picture the trio of craters Ptolemæus, Alphonsus and Arzachel appear to the right. Below and left of Ptolemæus is the ruined crater Hipparchus; above Hipparchus is Albategnius.

Above: This picture shows the Sinus Iridum, or Bay of Rainbows, which is bottom left. Most of the picture is covered by the huge Mare Imbrium. Photograph by M. Brown.

Above: Photograph of the Mare Nectaris, by M. Brown. At the top of the Mare is the great bay Fracastorius; the crater near the bottom right is Theophilus.

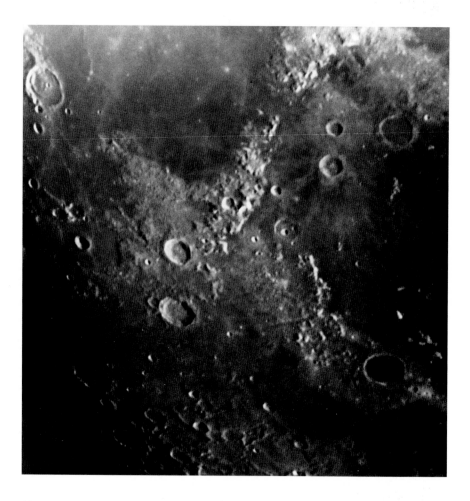

Above: Photograph of part of the lunar Apennines. These are the most impressive mountains on the whole of the Moon.

to the Jovian belts. Pickering believed this to be due to absorption in a lunar atmosphere, and he confirmed it on other occasions. He wrote that the dark band was seen only when the planet passed behind the Moon's bright limb; at the dark limb no such band was seen, and Pickering concluded that the atmosphere responsible for it was frozen out during the lunar night. Others who recorded the band were two of Pickering's colleagues, E. E. Barnard and A. E. Douglass, both of whom – Barnard particularly – were known to be expert observers.

From this kind of result, Pickering worked out a density of the lunar atmosphere as $\frac{1}{1,800}$ of that of the Earth's air at sea-level. This value was not only too great, but impossibly too great. Neither has the phenomenon been seen with any consistency, and it has never been photographed, which makes one suspicious; the human eye is very easily deceived.

Another line of investigation was tried out when the Moon occulted the Crab Nebula in Taurus, which is known to be the wreck of a star which exploded in a supernova explosion long ago, and which is a source of radio waves. (The supernova outburst was seen by Chinese and Japanese astronomers in the year 1054, but since the distance of the Crab is 6,000 light-years the actual explosion took place in prehistoric times.) Radio astronomers at Cambridge studied occultations of the Crab to look for effects roughly analogous to those for visual observations, and initially concluded that there were signs of a slight atmosphere round the Moon, but these results too were unconfirmed.

Neither are there any marked twilight effects on the Moon. Reports have been made from time to time, but proof is signally lacking. I may well be prejudiced here, since I have looked for twilight effects at the cusps of the crescent Moon more times than I can count, with completely negative results.

Two Russian astronomers, V. Fesenkov and Y. N. Lipski, attacked the problem in the 1940s. If there is any light coming from diffusion in a thin lunar atmosphere, it should have special qualities, and therefore be detectable with sensitive equipment. Fesenkov, in 1943, had no luck at all, and concluded that the Moon's atmosphere could not have a density greater than one-millionth of our own. Six years later Lipski made a new investigation, and obtained different results, announcing a value of $\frac{1}{20,000}$. This was confirmed in 1952, and in 1953 Lipski published another paper in which he raised the density to $\frac{1}{12,000}$ – agreeing well with earlier work carried out by Bernard Lyot in France.

Yet it was not long before doubts crept in. Lyot and his colleague Audouin Dollfus, working at the Pic du Midi Observatory in the Pyrenees, failed to confirm Lipski's results. After Lyot's death, Dollfus carried on the work and decided that the lunar atmosphere was too tenuous to be detected at all. This would mean a density not greater than one thousand-millionth of that of the Earth's air at sea-level. Clearly the lunar atmosphere was negligible by any standards, and would be very difficult to detect from Earth. Then, during the 1960s and 1970s, came the Apollo missions, and the first really reliable results were obtained.

Instruments carried in the orbiting sections of Apollos 15 and 16 detected small quantities of radon and polonium gas seeping out below the Moon's crust. This could be explained easily. Both gases are produced by the radioactive decay of uranium – and there is plenty of uranium in the lunar rocks. Finally, in 1972, came Apollo 17, carrying what was termed LACE – the Lunar Atmospheric Composition Experiment. The main gases proved to be helium and argon. Argon seeped out from below, and was at its greatest concentration a few hours before each lunar sunrise; the helium came from what is termed the solar wind, a stream of particles sent out by the Sun in all directions all the time.

Later, Drew Potter and Tim Morgan, using the 107-inch reflector at the McDonald Observatory in Texas, identified two more gases, potassium and sodium. The sodium appeared to surround the Moon rather in the manner of the coma of a comet. And there was an interesting development in 1998, at the time of the Leonid meteor shower.

Meteors, as we have noted, are cometary débris, and most of them are no larger than sand-grains. The Leonids, associated with the periodical comet Tempel-Tuttle, are seen every November, around the 17th of the month, but really rich displays occur only rarely – in general every 33 years, which is the orbital period of the comet itself.

There were good Leonid displays in 1799, 1833, 1866 and 1966 (1899 and 1933 were missed out, because the Earth did not pass through the main swarm). In 1998 the Leonids were back, peaking on 17 November. It seems that meteor impacts on the lunar surface created a cloud of sodium gas which then escaped from the Moon. Sodium atoms were 'pushed away' by the pressure of sunlight, and reached the vicinity of the Earth two days later. The cloud of lunar gas was 'gravitationally focused' into a narrow beam, which was photographed with a very sensitive camera at the McDonald Observatory.

What, then, about the possibility of seeing meteors in the Moon's atmosphere? With a ground density as high as the value given by Lipski, it could be possible. In 1952 I discussed the possibility with Dr E. J. Öpik of Armagh Observatory, a world authority on meteoritic phenomena. He then wrote as follows:

Lunar meteors are quite probable. Considering the surface gravity of the Moon, which leads to a six times slower decrease of atmospheric density with height, the length of path and duration of meteor trails on the Moon will be six times that on the Earth, if a lunar atmosphere about $\frac{1}{20,000}$ to $\frac{1}{100,000}$ of the density of the terrestrial atmosphere exists. At the same time, meteors of the size of fireballs will penetrate the lunar atmosphere and hit the ground. The average duration of a meteor trail on the Moon would be two to three seconds (as against half a second on Earth), because all meteors which can be observed on the Moon from such a distance must be large fireballs. The average length of the trail would be 75 miles, about $\frac{1}{30}$ of the Moon's diameter, so that meteors would be very short and slow objects.

Öpik added that with a 12-inch telescope it should be possible to record an average of one lunar meteor for every eight hours' observation.

All this seemed very reasonable at the time, and careful searches were made. I spent many hours at the eye-end of my main telescope, a 15-inch reflector, but the results were negative; I did not even suspect a single meteor. However, in the United States many streaks were reported by W. H. Haas and his colleagues in the Association of Lunar and Planetary Observers. The average trail-length worked out at about 75 miles, agreeing excellently with Öpik's estimate. Haas even calculated the probable diameter of an object he recorded in 1941; assuming it to have been a genuine lunar meteor it would have been 600 feet across – which, of course, means that it ought properly to be classed as a meteorite, not of cometary origin.

Now that we know the lunar atmosphere to be much too tenuous to produce meteoric trails, we must take a very critical look at these results. Whatever the observers saw, they can have recorded no meteors over the Moon, and meteorites would certainly be very rare indeed. The whole episode may be a warning that it is only too easy to 'see' what one half-expects to see (remember the canals of Mars!). There are only two possible explanations. Either the American observers were deceived by tricks of the eye, or else they were seeing phenomena of a completely different sort – which, in view of the descriptions given, seems most unlikely.

I also admit to being sceptical about a much more recent claim, made in November 1999. Observers in the United States reported seeing flashes on the unlit side of the Moon, which were attributed to Leonid meteors striking the surface. On 18 November, at 04.46.20s UT (Universal Time), Brian Cudnik was using a 14-inch reflector and recorded a flash, as bright as a fourth-magnitude star; a video made by David Dunham also recorded a flash. Other reports of flashes around the same time came in from elsewhere, and six were eventually listed, some of them as bright as the third magnitude. All the reports referred to the time when the Leonid shower was at its maximum. The main problem, to my mind, is that a sand-grain-sized meteor could not possibly produce a flash of this kind; it would need an object of much greater size. Meteorites are not associated with the Leonids or any other meteor shower; moreover, if there were exceptionally large Leonid meteors bombarding both Earth and Moon around this time we would expect them to produce brilliant fireballs – but this did not happen.

Over the years there have been occasional reports of isolated flashes, and meteoritic impact certainly cannot be ruled out. Probably the most convincing dates back to 15 April 1948. F. H. Thornton, a very skilled and experienced observer using an excellent 9-inch reflector, was looking at the dark-floored, 60-mile crater Plato when he saw something unusual:

> When I was examining Plato, I saw at its eastern* rim, just inside the wall, a minute but brilliant flash of light. The nearest approach to a description of this is to say that it resembled the flash of an AA shell exploding in the air at a distance of about ten miles. In colour it was on the orange side of yellow... My first thought was that it might be due to a large fall of rock, but I changed my opinion when I realised that close as it seemed to be to the mountain wall, it was possibly over half a mile away.

Was this a meteorite impact? It could have been. I have collected and analysed all the past reports I can find, but only Thornton's impresses me. Of course, the lunar atmosphere is no protection against meteorites – but neither is that of the Earth. The main difference is that our air is dense enough to burn up shooting-star meteors, while that of the Moon is not.

At least we now know the true situation. Take the whole of the lunar atmosphere, compress it to the density of the Earth's air at sea-level – and it will just about fill a cube with a side length of 200 feet. It may be regarded as a collisionless gas. The density is no more than one ten-thousand-millionth of that of our atmosphere, so that if we call the Moon 'an airless world' we are not far wrong.

* This was according to the old system. According to the IAU ruling, the flash occurred near Plato's western wall.

12

The Structure of the Moon

The Earth and the Moon have been closely linked ever since they were formed. Whether they once made up a single body (as most astronomers now believe) or whether they were formed separately, there can be no doubt about their association, and they are of around the same age. They have evolved differently, mainly because of the difference in mass; as we have already noted, the ratio is 81 to 1. The Moon lost its atmosphere at an early stage, and the main activity there died out a very long time ago. The Moon is also much less dense than the Earth, and it was realized well before the Space Age that there cannot be a heavy, iron-rich core comparable with that of the Earth. Moreover, there is no overall magnetic field today, though there may once have been.

Most of our detailed knowledge about the Moon has been due to space-research methods, beginning with the unmanned Russian Luna probes from 1959. I will discuss them later, so in a way I am now running ahead of my story, but it may be helpful to deal here with the internal structure of the Moon, together with 'moonquakes' and possible changes seen upon the lunar surface.

Without delving at all deeply into geology, I must introduce a few technical terms. Basalts are fine-graded, dark igneous rocks composed chiefly of material called plagioclase (feldspar) and pyroxene, together with other minerals such as olivine; breccias are made up of fragments of rocks cemented together by the effects of heat and pressure, while the material holding them together is called the matrix. Magma is molten rock below the surface, which becomes igneous rock when it solidifies, and is termed lava when it flows out on to the actual surface. On the Moon, particles larger than about one centimetre in diameter are called rocks, while smaller particles are known as fines.

The outer surface layer of the Moon is termed the regolith (fig. 30). It is a loose layer or débris blanket, continually churned by the impacts of micrometeorites; it is often referred to as 'soil', but this is misleading, because there is nothing organic about it. It is made up mainly of very small particles, with some larger rocks, a few feet in diameter, here and there. It is the regolith material which makes up the matrix of the breccias; it is rather variable in depth. In the maria it is around 6 to 25 feet deep, but it is thicker over the highlands, where it may in places go down to at least 30 feet.

En passant, there was a curious theory, due to Thomas Gold, that the maria at least might be covered with deep, soft, treacherous dust, into which a landing space-craft would sink. If this had been correct, then lunar travel would have been very difficult indeed, but it never really fitted the facts, and it was finally disproved in the 1960s, when unmanned probes came down on the surface unharmed. The regolith is firm enough to bear the weight of even a massive space-craft, and there seems little fear of finding any dust-drifts deep enough and soft enough to be dangerous.

Like the Earth, the Moon has proved to have a crust, a mantle and a core. One early finding was that both crust and mantle are thicker than with the Earth, no doubt because the temperatures deep inside the globe are so much lower. The highland crust averages just below 40 miles in depth, but is decidedly thicker on the far side of the Moon than on the Earth-turned hemisphere. The maria are volcanic, and cover 17 per cent of the surface, mainly on the near side; in general they are no more than 1½ miles deep. At a fairly shallow level there are areas of denser material, which have been located because an orbiting space-craft will speed up when passing over them; these are known as mascons (*mass con*centrations). They lie under the large basins, such as Imbrium and Orientale.

On the highlands, the rock fragments are mainly anorthosites, though there are other minerals as well, and the lunar landings have shown that the crustal composition varies from site to site. There is absolutely no evidence of 'hydrated' material – that is to say, material involving water in any form; I will come back to this later, when discussing the claim that here and there we could find ice.

For studying the Moon's interior we depend upon seismic methods – in fact, moonquakes – just as we depend upon earthquakes for information about the interior of the Earth. Of course they are very mild by our standards, and never exceed a value of 3 on the Richter scale, so that an

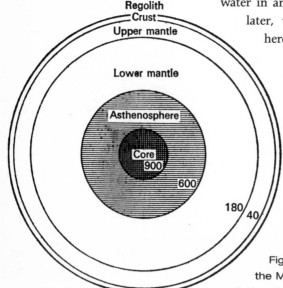

Fig. 30. A cross-section through the Moon (figures indicate miles below the surface)

observer standing right above the focal point of a moonquake would hardly be aware of it; there is no danger on this score to a future Lunar Base (in fact, the Moon is much safer seismically than Tokyo or San Francisco).

Moonquakes are of two main kinds (fig. 31). Most originate from a zone 500 to 600 miles below the surface, and are common enough. Shallow moonquakes also occur, at depths of from 30 to 125 miles, but are rather less common. All in all, the seismographs set up by the Apollo astronauts recorded 1,300 moonquakes per year (meteorite impacts were also recorded, but luckily these produce different types of wave-patterns, so that they can be weeded out from the true moonquakes). Some of the quakes seem to be triggered off by the tidal pulls of the Earth and the

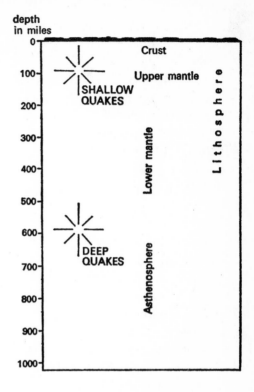

Fig. 31. Depths of moonquakes

Sun, while others could be caused by the great temperature changes between day and night, causing the rocks to expand and contract.

There have also been man-made moonquakes, caused by the impacts of discarded lunar modules. These show that the outer few miles of the Moon are made up of cracked and shattered rock, so that the signals can echo to and fro. It was even said that after the impact of an Apollo module the Moon 'rang like a bell'.

Below the crust comes the mantle, whose structure seems to be relatively uniform. The seismic effects of a 1-ton meteorite which hit the Moon in July 1972, and was recorded by an Apollo seismograph, indicated that from 600 to 800 miles below the surface, in the region termed the asthenosphere, the rocks are hot enough to be molten (earlier Apollo measurements had disposed of a theory that the Moon might be cold and solid all through its globe).

Finally, there seems to be a metallic core with a diameter of no more than 600 miles. It contains iron, but is much smaller than the Earth's core, not only absolutely but also relatively; neither is it so hot, and can hardly exceed 1,500 degrees C. If there were an iron-rich core, then we would expect a magnetic field, but in fact the overall field is negligible. On the other hand, rocks brought back to Earth have shown that there are some locally magnetized regions, and it may well be that in the distant past there was a general field, which has now died out for reasons which remain unclear.

Quiescent though it is, the Moon is not completely inert. Very mild activity is seen occasionally, in the form of localized glows and obscurations. These events are known generally as TLP or Transient Lunar Phenomena – a term for which I believe I was originally responsible. They are very mild indeed, and until recently there was a great deal of scepticism about them, mainly because most of the reports – not all – came from the work of amateur astronomers.

There were, however, some professional observations. Thus in 1892 E. E. Barnard (discoverer of Amalthea, the fifth satellite of Jupiter, as well as a number of comets) saw the bright ray-crater Thales filled with 'pale luminous haze', though the surrounding features were absolutely sharp and clear-cut. Once, in 1902, the French astronomer Charbonneaux, using one of the world's largest refractors (the Meudon 33-inch, at the Paris Observatory), described how he saw a small but unmistakable white 'cloud' form close to Theætetus, in the region of the Apennines, and various localized obscurations were reported by W. H. Pickering.

Variations were reported, too, inside the 60-mile, dark-floored Plato, one of the most studied formations on the Moon. The craterlets on the floor are sometimes invisible when they should be obvious. I can cite a personal case here. Before midnight on 3 April 1952 I was unable to see them at all, though I too was using the Meudon refractor under good conditions; four hours later T. A. Cragg, now of the Mount Wilson Observatory, looked at Plato with a 12½-inch reflector and saw that the floor looked blank. He was in no doubt that some local obscuration was responsible. Plato has a long history of similar anomalies.

Despite the professional observations, it was true that the bulk of the reports came from amateurs, and the local obscurations were officially dismissed as being due to imagination or tricks of the eye. Those who believed otherwise, such as myself, were not taken very seriously. This was understandable enough, because to see one genuine event means many hours of fruitless checking – something which no professional, busy with more important matters, has time to do. So the amateurs, notably those of the Lunar Section of the British Astronomical Association, went on with their patient work and waited to see what would emerge from it.

The whole situation was transformed by one episode, in which I played a minor and totally undistinguished part. During 1955 Dr Dinsmore Alter, using the 69-inch reflector at Mount Wilson, took some photographs of the two large walled plains Alphonsus and Arzachel, which are members of the

Ptolemæus chain, and which lie near the centre of the Moon's disk as seen from Earth. Alter took pictures in both infra-red light and in blue-violet. As almost everyone knows, infra-red will penetrate haze or mist, while light of shorter wavelength will be blocked or scattered by it. (This is why our sky is blue; the shorter wavelengths of the Sun's radiation are spread around, while the longer wavelengths are not – at least, not to the same extent.) Alter naturally expected that in his sets of photographs Alphonsus and Arzachel would be similarly clear-cut, but he found that on several occasions part of the floor of Alphonsus was blurred in the blue-violet pictures. He wrote: 'For some reason the blue-violet photographs lose more detail in the east side of Alphonsus than they do in the floor of Arzachel. This is not true of the infra-red ones... There is a temptation to interpret these results immediately as being due to a thin atmosphere, either temporary or permanent, over the floor of Alphonsus. The theoretical difficulties inherent in such a hypothesis are, however, strong enough to forbid whole-hearted acceptance of it.'

This was by no means the first time that local effects had been reported in Alphonsus. I had some correspondence with Alter and various other professional astronomers, including N. A. Kozyrev, of the Crimean Astrophysical Observatory in the USSR. I suggested that it would be worth while to keep a careful watch on the area, and Kozyrev was among those who did so, using the 50-inch reflector at the Crimea. His method was to take regular spectrograms (that is to say, photographic spectra), and since the Crimean reflector has no separate guiding telescope he had to watch during the exposure time to make sure that there was no drift of the image. While doing this, at 0100 hours GMT on 3 November 1958, he noticed that the central peak of Alphonsus had become blurred, and was apparently engulfed in a reddish 'cloud'. Another spectrogram, taken between 0300 and 0330 hours GMT, proved to be remarkably interesting. While guiding the telescope, Kozyrev kept his eyes on Alphonsus, and noticed that the central peak had become abnormally bright. Suddenly the brilliance began to fade; Kozyrev immediately stopped the exposure and started a new one, which was completed at 0345 GMT. By the time it was finished, everything was normal once more, and Alphonsus looked the same as it usually does.

The announcement of Kozyrev's results took many astronomers by surprise. Kozyrev himself was quite definite: 'On the spectrogram recorded on 3 November, 1 hour UT, the central peak of the crater appears redder than normal. Probably at this time the peak was being observed and illuminated by the Sun through the dust (ashes) being thrown up by the eruption.' In a letter to me written shortly afterwards, he said that the spectrograms showed that 'hot carbon gas had been sent out, causing a rise in temperature of perhaps 2000 degrees'.

I am frankly sceptical about this. Such a rise in temperature seems improbable, to put it mildly, but at least it seemed that an event of some sort had taken place – and on checking the records, I found that as long ago as 1882 the German astronomer Klein had claimed that he had seen 'volcanic phenomena' inside Alphonsus. On the other hand, it is very dangerous to put a great deal of faith in old records of this kind.

The next step was to see if there had been any permanent change in the area. During the following weeks and months several observers reported red patches near the site, and these were interpreted as being due to coloured material thrown out at the time of the disturbance; for instance Brian Warner, now Professor of Astronomy at the University of Cape Town but then working at the University of London Observatory, used the 18-inch refractor there, and described the patch as 'bright red'. I was less successful, and I admit that I was never able to see any red patch at all; it certainly seems to be absent now, but the 1958–9 reports are not at all easy to explain away.

Since then there have been other records of red TLP in the area, plus one more spectrographic observation by Kozyrev on 23 October 1959, though on that occasion nothing unusual was seen visually.

The next major development came in 1963 when on 30 October, J. Greenacre and E. Barr, at the Lowell Observatory in Arizona, observed colour in the region of Aristarchus. The phenomena included red and pink patches, and were quite unmistakable. In the following month similar events were seen, and were confirmed by P. Boyce at the Perkins Observatory, using a 69-inch reflector.

Aristarchus is the brightest crater on the Moon. It is only 23 miles in diameter and 6,000 feet deep, but it is so brilliant that it is always instantly recognizable even when lit only by earthshine; in 1787 no less a person than Sir William Herschel mistook it for a volcano in eruption. It has a central mountain; the walls are terraced, and are crossed by dark bands which are easy objects to see. The bands are due to differences in level and in surface texture, as we now know from the Orbiter and Apollo photographs. Close beside Aristarchus is Herodotus, of similar size but with a darkish floor, and from Herodotus extends the great valley often known as Schröter's Valley in honour of its discoverer. Aristarchus is the most event-prone crater on the Moon, and it is responsible for more than half the total number of TLP. Gaseous emissions from it have been confirmed spectrographically, and on 19 July 1969 activity was seen by the astronauts of Apollo 11, who were then together in the Command Module orbiting the Moon. Armstrong, Aldrin and Collins used binoculars and reported a luminous north-west wall, 'more active' than anywhere else on the surface.

Added confirmation came from observers on Earth, who reported TLP in Aristarchus at and around that time.*

To run ahead of the story: there were further developments with the flight of Apollo 15, in 1971, when the command module carried a special device intended to detect what are known as alpha particles. These are produced by the decay of the radioactive gas radon, which in turn comes from uranium and thorium. If radon gas diffuses through the regolith, atoms will be released; when these decay, alpha particles will be emitted. This is precisely what was found; as Apollo 15 passed 70 miles above Aristarchus, there was a significant rise in the numbers of alpha particles emitted by radon-222 (that is to say, the radon isotope with an atomic weight of 222). Particles associated with the decay of the other main kind of radon, radon-220, were not found, and this showed that the effect was not due merely to a local excess on the surface of uranium and thorium. The radon isotopes must have diffused through the regolith from below. Radon-222 could be (and was) detected, but radon-220 atoms have a very brief existence; they last for less than one minute, and would not persist for long enough to come through the regolith. Therefore, there seems no doubt that radioactivity in the Aristarchus area is responsible, and the scientists who carried out the analysis – P. Gorenstein and P. Bjorkholm – added that 'The observed radon emanation is associated with the same internal processes which will on occasion emit volatiles in sufficient quantity to produce observable optical effects.'

Another interesting report came on 23 May 1985, from the Greek observer G. Kolovos and his colleagues. Near a small, irregular crater in the Palus Somnii they photographed a brief TLP covering an area of $3\frac{1}{2}$ x 3 miles, and attributed it to gases sent out from below the crust. Attempts were made to explain this event away by an artificial satellite, reflecting the sunlight, which just happened to be in the line of fire, but the photographs indicated that the obscuration was linked with the topography of the area, so that the satellite theory is not at all convincing.

Over the years I have recorded a number of TLP – most notably on 30 April 1966, in the area of the walled plain Gassendi. This was the most unmistakable event I have seen, and it was also recorded by several independent observers at different sites. The main feature was a wedge-shaped, reddish-orange streak extending from the wall of Gassendi right across to the central peak. On that occasion I was using my 15-inch reflector under excellent conditions of seeing.

* Unfortunately I could take no part in the observational programme; I was in the BBC television studio throughout the mission, carrying out live commentaries.

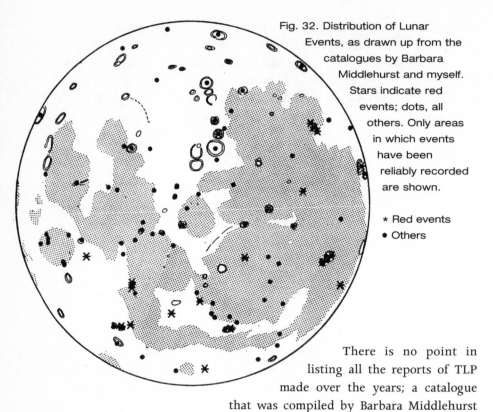

Fig. 32. Distribution of Lunar Events, as drawn up from the catalogues by Barbara Middlehurst and myself. Stars indicate red events; dots, all others. Only areas in which events have been reliably recorded are shown.

⋆ Red events
• Others

There is no point in listing all the reports of TLP made over the years; a catalogue that was compiled by Barbara Middlehurst and myself, published by NASA, includes hundreds (fig. 32). Of course, many of these are certainly spurious; it is only too easy to be deceived by conditions in our own atmosphere. It was very reasonable for most professional astronomers to remain unconvinced, but finally a report from the eminent French astronomer Audouin Dollfus, published in 1999, seemed to settle the problem. Using the Meudon telescope, Dollfus wrote as follows:

> On 30 December 1992, glows have been recorded at the lunar surface, on the floor of the crater Langrenus. They were not present the day before. Their shape and brightness were considerably modified three days later. These glows appeared briefly also in polarised light. They are apparently due to dust grain levitations above the lunar surface, under the effect of gas escaping from the soil. The Moon appears as a celestial body which is not totally dead.

It seems that most TLP are seen near the boundaries of the regular seas, and in areas which are rich in rills. They also seem to be commonest near lunar perigee, when the Moon is at its closest to the Earth and the surface is under maximum strain. Moreover, there is a definite correlation between TLP sites and the genuine moonquakes. Everything seems to fit. It is fair to say that amateur observers (such as myself) were elated by the Dollfus report; not for the first time, we had been shown to be right.

When we come to observable structural changes on the Moon, the situation is very different. There have been various reports of definite alterations. Frankly, I have no faith in any changes of this kind; but it is always interesting to delve into history, and a good place to begin is at Linné, on the Mare Serenitatis. Linné is one of the most studied objects on the whole Moon, and observers have every reason to be grateful to it, since it was the direct cause of the reawakening of interest in selenography from 1866 onward.

Lohrmann, in 1834, looked at Linné and described it as 'the second most conspicuous crater on the plain... it has a diameter of about 6 miles,* is very deep and can be seen under all angles of illumination'. Mädler, at about the same time, wrote: 'The deepness of the crater must be considerable, for I have found an interior shadow when the Sun has attained 30 degrees. I have never seen a central mountain on the floor.' Both observers drew it, measured it and used it as a reference point. It also appears as a conspicuous crater on six drawings made by Julius Schmidt between 1841 and 1843.

All this seemed definite enough. Yet on 16 October 1866, Schmidt was examining the Mare Serenitatis when he suddenly realized that Linné had disappeared. Where the old crater had stood, all that remained was a small whitish patch. It was a startling discovery, but Schmidt had no doubts about it.

His announcement caused a tremendous sensation. Up to then, Mädler's view of the Moon as a dead, changeless world had been accepted without question for many years, and astronomers were not inclined to change their opinions. Hundreds of telescopes were pointed at Linné, and many drawings were made of it (photography was still at an early stage). The results were not in good agreement, but at least it was clear that the deep crater described by the old observers had gone – if, of course, it had ever existed. In its place was a whitish patch, containing a tiny feature which was sometimes described as a craterlet and sometimes as a hill. One particularly famous observer, Angelo Secchi – a pioneer of stellar spectroscopy – looked at it on 11 February 1867, using the powerful telescope at the Vatican Observatory, and wrote that 'there is no doubt that a change has occurred'. Sir John Herschel suggested that a moonquake had shaken down the walls of the old crater, and that the hollow had been filled with rising lava.

---

* Lohrmann actually said 'somewhat more than 1 mile', but the old German mile is equal to 4½ of ours. Mädler gave 1.4 German miles, thus agreeing with Lohrmann.

Probably the leading active selenographer of the time was Edmund Neison, author of the classic book published in 1875. I have found an article by him written for the *Quarterly Journal of Science* for January 1877, and it seems worth quoting:

> *According to three or more independent selenographers, the most experienced and eminent that Science has seen, the object named Linné was a conspicuous crater of large diameter and depth. Now in its place all that exists is a tract of uneven ground, containing a small, scarcely-visible, insignificant crater-like object. It is impossible that the one could ever be systematically mistaken for the other. It is inconceivable how our three greatest selenographers could have systematically and independently made the same blunder, and that one blunder only... A real physical change on the Moon's surface must therefore have occurred at this point.*

Of the 'three greatest selenographers' (Lohrmann, Beer and Mädler) only Mädler remained alive in 1866. He accepted the evidence of change, but he nevertheless wrote that when he observed Linné in May 1867 he 'found it shaped exactly, and with the same throw of shadow, as I remember to have seen it in 1831. The event, of whatever nature it may have been, must have passed away without leaving any trace observable by me.'

Linné certainly shows apparent changes due to the angle of illumination. We now know it to be a small, clear-cut impact crater, surrounded by a bright area. There seems no doubt that it has always looked the same as it does now.

We can also discount other alleged structural changes, but it may be worth mentioning them briefly for purely historical reasons. For example, Schröter drew a large distinct crater with bright walls and a dusky floor on the border of the Mare Crisium, and named it Alhazen; Mädler could not find it, and transferred the name to another formation. Mädler himself showed a 5-mile crater near Alpetragius which had similarly 'gone missing' when Schmidt looked for it. The large walled plain Cleomedes contains a small crater considered by Schröter to have been formed in October 1789, since he had missed it previously and saw it clearly afterwards. On 27 May 1878 the German observer H. Klein recorded Hyginus N, a rimless depression close to the famous rill, which he said was 3 miles in diameter and filled with shadow under oblique lighting; he had not seen it before, and thought that it must be new. Most similar reports date back to the nineteenth century, and must be rejected out of hand. Remember that before 1866 (the Linné period) there were very few observers looking seriously at the Moon, and Mädler, the best of them, used a telescope of less than 4 inches aperture. But one final case may be of interest.

On the Mare Fœcunditatis there are two small craters, Messier and Messier A, from which extend a double ray which gives the pair a strange resemblance to a comet. Beer and Mädler made over three hundred drawings of them, between 1829 and 1837, and described them as being exactly alike. 'To the west of Messier there appears an identical formation. Diameter, shape, height and depth, colour of interior are the same, and even the positions of the peaks; everything points to the fact that we have here either a most remarkable coincidence, or that some as yet unknown law of nature has been at work.' Today, Messier and Messier A generally appear different in both shape and in size, A being the larger and less regular; but the angle of sunlight striking them is all-important here, and there is no doubt at all that the alterations are purely optical. Under some illuminations they really do look alike. (H. H. Nininger, the American meteorite expert, suggested in 1952 that Messier and its twin had been formed by a meteorite which ploughed its way through a ridge!)

There is always the chance that a major meteoric impact will produce a new visible crater on the Moon, and if so we could well track it down, because maps of the lunar surface are now so detailed. But internal activity strong enough to change the shape or form of a crater died away long ago, and major alterations belong to the remote past. If we could project ourselves back to the time of the dinosaurs and turn a telescope toward the Moon we would see the mountains, the valleys, the seas and the craters looking to all intents and purposes exactly as they do today.

13

Eclipses of the Moon

A total eclipse of the Sun is the grandest sight in all Nature. As the last segment of the brilliant solar disk is covered up, the corona flashes into view, together with red prominences; the sky darkens, and planets and bright stars shine forth. Unfortunately there are many people who have never seen a total eclipse, because one has to be in just the right place at just the right time – and, moreover, under a cloud-free sky.

Since the Moon appears only just large enough to cover the Sun completely, the belt of totality can never be more than 169 miles wide, and is generally much less. To either side of this narrow zone all that can be seen is an unexciting partial eclipse. For example, the eclipse of 30 June 1973 was total over parts of Africa, but from the south coast of England there was only a small 'bite' out of the Sun, and over the rest of Britain there was no eclipse at all. The last total solar eclipse visible from any part of England was that of 11 August 1999; the track crossed Cornwall and parts of Devon, but in most places the weather was poor (from Falmouth in Cornwall, right on the central line, I sat disconsolately under an umbrella, listening to the rain). The next English eclipse will be that of 23 September 2090, which I will not see – unless, of course, I live to the advanced age of a hundred and sixty-seven.

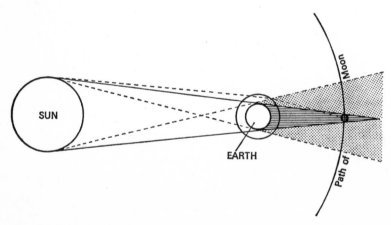

Fig. 33. A lunar eclipse

With an eclipse of the Moon, the whole situation is different (fig. 33). The eclipse is visible from a complete hemisphere of the Earth, because it is due to the Earth's shadow and not to any solid body blocking out the lunar disk. If a lunar eclipse is due, and the Moon happens to be above the horizon from your observing site, you will see it.

The result is, of course, that from any particular point on the surface of the Earth eclipses of the Moon are much commoner than eclipses of the Sun. On the other hand they are much less spectacular, and not nearly so important. All that really happens, from the viewpoint of the casual observer, is that the Moon becomes dim and changes colour during its passage through the shadow.

As fig. 34 shows, the principal cone of shadow cast by the Earth, know as the umbra, is very long. Its average length is 850,000 miles, which is more than three times the distance of the Moon from the Earth, so that at the mean distance of the Moon (239,000 miles) the cone has a diameter of about 5,700 miles. In the second diagram I have tried to show this to a more accurate scale. In any case, it is clear that the umbra is large enough to cover the Moon completely, and totality may last for as much as an hour and three-quarters, remembering that in the course of an hour the Moon moves across the sky by an amount slightly greater than its own apparent diameter.

Every scrap of direct sunlight is cut off from the Moon as soon as it passes into the umbra, and it might be thought that the Moon would simply vanish. This does not happen, because a certain amount of sunlight is refracted on to its surface by the Earth's atmosphere. In fig. 33, one of these refracted rays is shown by the dotted line, and clearly it strikes the Moon, even though the Moon itself is directly behind the Earth. Instead of disappearing completely, the Moon becomes dim, and often shows strange and beautiful colour effects. Lunar eclipses may not be important, but they are undeniably lovely. During totality the effects of stars shining out near the dimmed Moon are particularly impressive.

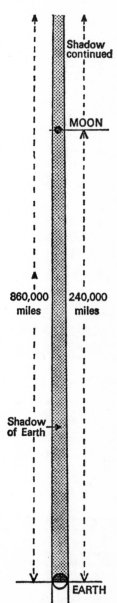

Fig. 34. Mean distance of Moon
in relation to Earth's umbra

Because the Sun is a disk, and not a mere point of light, the umbra of the Earth's shadow is bordered by what is termed the penumbra. This, too, causes a dimming of the Moon, but the effect is not nearly so marked. Since the Moon has to pass through the penumbra before it enters the umbra, there is a slight falling-off of light to one side of the Moon before the main eclipse begins. Many books claim that the penumbra is undetectable except by a skilled observer under good conditions, but I disagree strongly; I have never had the slightest difficulty in seeing the penumbra well upon schedule.

Not all lunar eclipses are total. Sometimes they are partial, and on other occasions penumbral, when the Moon avoids the main shadow-cone altogether. Obviously, an eclipse can take place only when full moon occurs near a node. On average, at least one lunar eclipse can be seen from any point on Earth each year, but not all of them are total. For example, there were no total eclipses in 1998 or 1999, though on 28 July 1999 there was a partial, with 40 per cent of the Moon covered. There were two totalities in 2000 (21 January and 16 July), but only the first of these was visible in Britain. A list of recent and future eclipses is given in Appendix 3.

Eclipses may be predicted for many years ahead, because the movements of the Sun, Earth and Moon are so accurately known. As long ago as 600 BC Thales of Miletus, first of the great Greek astronomers, was able to forecast eclipses fairly well by using a much rougher method. He knew that the Sun, Moon and node return to almost the same relative positions after a period of 18 years 10¼ days, the so-called Saros, so that any solar or lunar eclipse is apt to be followed by another eclipse 18 years 10¼ days later. (This is according to our modern calendar, and allows for five leap-years in the meantime.) The method is only approximate, because the relative positions are not exactly the same, but it is better than nothing. For instance, there was a total lunar eclipse on 29 January 1953, visible from England. Adding one complete Saros period brings us to 10 February 1971 – and sure enough there was another total eclipse; but this time it was only partly visible from England, because the Moon set before the eclipse was over.

By using the Saros method, Thales was able to make some reasonable predictions, even though he had no idea of how eclipses are caused. In those early times it was not even known that the Earth is a globe. Yet by 450 BC Anaxagoras of Clazomenæ was well aware of what happened, and he also reasoned that because the Earth's shadow on the Moon is curved, the earth itself must be spherical.

Eclipse records go back almost as far as written history itself, but most refer to the Sun, and the oldest lunar eclipse on record seems to have been that observed by the Chinese in 1135 BC. Two later eclipses are worth mentioning here, because they have a place in history, and one of them even had a marked influence upon the whole sequence of events in the Classical period.

In 413 BC the Peloponnesian War was raging; the two chief Greek states, Athens and Sparta, were fighting for supremacy, and the Athenian army which had invaded Sicily was in serious trouble. In fact things were so bad that Nicias, the Athenian commander, decided to evacuate the island altogether, and if he had done so at once all might have been well. Unfortunately there was a total lunar eclipse on the night before the evacuation was due to begin, and Nicias believed that it had been sent as a warning. The astrologers were called in, and advised that the army should stay where it was 'for thrice nine days'. Nothing could have suited Gylippus, the enemy commander, better. He attacked the waiting Athenian fleet, destroyed most of it and blockaded the rest in harbour. The trapped Athenian army was utterly wiped out; Nicias lost his life, and eight years later Athens lay at the mercy of Sparta.

The story of the 1504 eclipse is not only more modern, but also more cheerful. At that time Christopher Columbus was in the island of Jamaica, and difficulties arose when the local inhabitants refused to supply him and his men with food. Unlike Nicias, Columbus knew a great deal about lunar eclipses, and he remembered that one was due on 1 March, a day or two ahead. He therefore told the Jamaicans that unless they mended their ways he would make the Moon 'change her colour, and lose her light'. The eclipse duly took place, so alarming the natives that they immediately raised Columbus to the rank of a god. No further trouble was experienced with food supplies!

The effect on the Jamaicans would have been even greater if the Moon had disappeared completely, as had been known to happen. There were two total eclipses in 1620, the first of which was watched by Kepler, and each time it seems that the Moon became utterly invisible; Hevelius noted the same thing in 1642, while in 1761 the Swedish astronomer Per Wargentin (after whom the famous lunar plateau is named) observed an eclipse in which the Moon vanished so thoroughly that it could not be found even with a telescope, though nearby faint stars shone out quite normally. Beer and Mädler saw a very dark eclipse in 1816, and in 1884 the shadowed Moon could only just be made out. On the other hand, the eclipse of 1848 was so bright that it was hard to tell that an eclipse was in progress at all, apart from the fact that the Moon turned a curious shade of blood-red.

I have seen a good many lunar eclipses now, and no two are alike. I particularly remember the eclipse of 29 January 1953; the predominant colour was coppery pink, with some glorious bluish hues and also what can only be termed flame-colour. The eclipses of 30 December 1963 and 10

December 1973 were unusually dark, though the eclipsed part of the Moon never vanished completely. On 24 April 1986 there was a total eclipse, not seen from Britain but well placed from Bali, in Indonesia, where I was at the time. This was a fairly bright eclipse, but the sky became dark enough for stars to be seen – and there too was Halley's Comet, now fading as it moved away from the Sun. The eclipse provided the very last opportunity to see the comet with the naked eye, at least until it comes back once more in 2061. On 15 June 1992 I saw a total eclipse from Florida, and this time the Moon really did become dark; the eclipse was not total – 68 per cent of the surface was in shadow – but using binoculars I could not see the eclipsed portion at all, and I learned that others were equally unsuccessful even with telescopes. On the other hand, at the total eclipse of 27 September 1996 the whole of the shadowed area remained bright orange, and detail could be seen in it.

Conditions in the Earth's air are responsible for these variations, because all the light which reaches the eclipsed Moon has to be refracted through our atmosphere. Various correlations have been found. The tremendous volcanic explosion of Krakatoa in 1883 scattered so much dust in the upper atmosphere that its effects were traceable for months afterwards, and probably caused the darkness of the 1884 eclipse. Other dark eclipses can be similarly linked; that of 1902 with the eruption of Mont Pelée, that of 1950 with vast forest fires raging in Canada, that of 1963 with the outbreak at Mount Agung in the East Indies, and so on. But this may not be the whole story.

The French astronomer A. Danjon (whom I knew well; he died some years ago) introduced an 'eclipse scale' which is distinctly useful. On this system, 0 indicates a very dark eclipse, with the Moon almost invisible near mid-totality; 1, dark eclipse with greys and browns, and with details on the surface barely identifiable; 2, deep red or rusty, with the outer edge of the umbra relatively bright; 3, brick-red, with a bright or yellow rim to the shadow; 4, very bright, orange or coppery red, with a bright bluish shadow rim. Danjon also tried to link the brightness of a lunar eclipse with conditions on the Sun. It has long been known that the Sun has a reasonably regular cycle of activity, with spot-maxima occurring every eleven years (in 1989–90 for instance, and again in 2001), while near minimum activity the solar disk may remain spotless for many days consecutively. According to Danjon, eclipses of the Moon are dark for the two years after solar maximum; they then become brighter until the seventh or eighth year, when they reach 4 on his scale; subsequently there is an abrupt decrease. My own records do not really support this idea; I believe that if there is any real connection the effects must be very minor.

Another idea was that the luminescence of the lunar rocks themselves could be involved, but calculations have shown that luminescent effects are too slight to be detectable.

To the selenographer, the most important fact about a lunar eclipse is that there is an abrupt cut-off of sunlight. Since the Moon is virtually without atmosphere, and the surface rocks are very poor at retaining heat, a sudden wave of cold sweeps across the Moon. During the 1939 eclipse, Pettit and Nicolson at the Mount Wilson Observatory found that the temperature dropped from +160 degrees F to -110 degrees F in only an hour, and this sort of result has been confirmed since. (The temperatures measured at radio wavelengths do not show the same variation, because here we are dealing with regions slightly below the Moon's surface, and the outer rocks are excellent insulators.)

Not all parts of the Moon cool down at the same rate. Infra-red techniques have shown that during an eclipse there are some regions which cool down relatively slowly, and these regions have become known as 'hot spots'. Actually, the term is misleading; all it means is that when the Moon is in the shadow of the Earth, these spots are less chilly than their surroundings. Over four hundred such areas have been located, some of them associated with major craters. The best example is Tycho, centre of the bright ray system, in the southern part of the Moon.

There is no chance that internal heat from below the crust can play a rôle, and no doubt the surface texture in these regions is responsible. It is significant that two of the main 'hot spots' are Tycho and Copernicus, the two most important ray-centres on the Moon; others are Aristarchus, the brightest of all the craters, and various smaller ray-systems and bright craterlets, though not all the 'hot spots' are associated with rays.

There have been suggestions that the sudden wave of cold during an eclipse may cause detectable changes in various features. Eighty years ago W. H. Pickering, in America, reported changes in Linné which he put down to the precipitation of snow or hoar-frost; he believed that during an eclipse the white patch increased in size, to decrease again when the eclipse ended and the temperature returned to normal. But in Pickering's time the lunar atmosphere was thought to be much denser than we now know it to be, and any sort of precipitation is out of the question. I have made careful measurements at several eclipses, with negative results. Neither is there any confirmation of eclipse-induced changes in the dark-floored plains such as Plato and Grimaldi.

There is no glare from the full moon during a total eclipse, and occultations of stars can be well seen; unfortunately bright stars are seldom in the right place at the right time, and the only recorded instance of a bright planet (Jupiter) being occulted by a totally eclipsed Moon dates back as far as the year 755.

A lunar eclipse is not so exciting as a total eclipse of the Sun; there are no prominences or coronal rays, and everything happens much more slowly. Yet the passage of the Moon through the dark cone of shadow cast by our own world has a quiet fascination all its own.

14

The Way to the Moon

The idea of travel to the Moon is by no means new. In the second century AD a Greek satirist, named Lucian, wrote a story called the *True History*, which described a lunar voyage carried out quite involuntarily. According to Lucian, a ship carrying a full crew through the Straits of Gibraltar was caught in a waterspout, and was hurled upward for seven days and seven nights until it landed on the Moon. Lucian did not mean to be taken seriously – he admitted that his 'true history' was nothing but lies from beginning to end – but even in those far-off times it was known that the Moon, like the Earth, had a surface covered with mountains, valleys and plains.

Various other extravagant suggestions were made during the period following the invention of the telescope. In Kepler's classic *Somnium*, the hero was carried moonward by demon power. Four years later, in 1638, came *The Man in the Moone*, written by an English bishop, Francis Godwin, in which the astronaut was flown to the Moon on a raft pulled along by wild geese – to be greeted by a race of giants talking in a language so musical that it could be written down only in note form. Lunar society, wrote Godwin, was distinctly puritan. Any Moon-child showing signs of latent wickedness was promptly despatched to the Earth, where there is already so much evil that a little more will not matter. (Today, looking round the world of AD 2000, who can say that Godwin was wide of the mark?)

Godwin, like Lucian, was writing with his tongue very much in his cheek, but another English bishop of the same period, John Wilkins, was not. In his *The Discovery of a World in the Moone*, Wilkins maintained that the Moon is inhabited, and suggested that the British Government should annex it for the nation. The interesting thing about this book is that the author meant it to be taken very seriously indeed, and by the standards of his time he was no eccentric; he later became Secretary of the Royal Society.

Passing over Cyrano de Bergerac's *Voyages to the Moon and Sun*, in which the space-travel methods ranged from sucked-up dew to exploding fire-crackers, we must come forward to 1865, which is an important date in lunar history inasmuch as it saw the publication of one of the most famous science-fiction stories of all time, *From the Earth to the Moon* by Jules Verne (followed a few years later by its sequel, *Round the Moon*). It introduced the idea of a space-gun, in which the travellers were sent to their target inside a projectile fired from a huge cannon. Verne was not himself a scientist, but he believed in checking his facts, and his novel makes splendid reading even today. Obviously, the whole concept of the space-gun is outmoded, but Verne chose the correct escape velocity (7 miles per second), his fictional launching-ground was not far from the modern Cape Canaveral, and the giant telescope he described on Long's Peak echoes the real Hale 200-inch reflector on Palomar Mountain. It is, I feel, significant that when the Russians sent their first camera-carrying vehicle on a trip round the Moon, in 1959, one of the far-side craters that they discovered was promptly named in honour of Jules Verne.

There are many books dealing with the history of space research, and I do not propose to attempt it here. So I will do no more than say that the only valid method of sending vehicles to the Moon is by rocket power, since this involves the principle of reaction – 'every action has an equal and opposite reaction' – and there is no need for a surrounding atmosphere, as there is with an ordinary aircraft. In a Guy Fawkes rocket, for instance, the hot gas rushing out of the exhaust propels the tube of the rocket in the opposite direction; the rocket kicks against itself, so to speak, and is at its best in vacuum, where there is no resisting atmosphere to be pushed out of the way. The principle of a modern space-vehicle is the same, even though the solid gunpowder of the firework is replaced by a tremendously complicated rocket motor powered by liquid propellants. One man who worked out the main theory in considerable detail, even before the end of the nineteenth century, was the Russian schoolmaster Konstantin Eduardovich Tsiolkovskii, who published some articles about it in obscure journals which caused absolutely no comment – for the simple reason that almost nobody knew about them. It was only toward the end of his life that Tsiolkovskii achieved fame. He never built a rocket, and was in no position to try; but I recommend you to read his novel *Beyond the Planet Earth*, written in 1896. As a story, and as a literary exercise, it can only be described as atrocious, but as a forecast it was decades ahead of its time.

Tsiolkovskii knew that liquid fuels would have to be used instead of the weak, uncontrollable solids of the gunpowder variety. In 1926 Robert Hutchings Goddard, in America, actually fired the first liquid-propelled rocket in history, managing an altitude of 184 feet and a top speed of 60 mph. Again there was no general comment, because Goddard was not interested in publicity; but between that modest beginning and the start of the war, thirteen years later, there were spectacular developments, mainly in Germany. An initially amateur team was taken over lock, stock and barrel by the

German Government, and on Hitler's orders a research base was set up on the Baltic island of Peenemünde. There, a team led by Wernher von Braun developed the V2 weapon – a rocket, and the direct ancestor of the space-craft of today.

At the end of the war, the centre of interest shifted to America. Rockets were improved year by year, and even the most diehard opponents of space-travel began to realize that the Moon was coming within reach. Yet before long the initiative passed to the Soviet Union and it is my firm view that the real start of the Space Age can be fixed as 4 October 1957, when the Russians launched their first artificial satellite – Sputnik 1, which sped blithely round the world sending back its 'Bleep! Bleep!' radio signals which sounded, to some people, faintly derisory. In the following year the American team led by von Braun managed to send up a satellite, and also did their best to despatch vehicles to the Moon. Four launches ended in failure, and again the Russians achieved a notable 'first'.

Strictly speaking there were three 'firsts', all in 1959 (fig. 35). In January the probe Luna 1 passed within 4,660 miles of the Moon, proving, among other things, that there is no appreciable lunar magnetic field. In September Luna 2 made a crash-landing on the lunar surface, thereby forging the first direct link between our world and another; and on 4 October – exactly two years after the ascent of Sputnik 1 – Luna 3 started on a journey which took it right round the Moon, giving us our first positive information about those parts of the surface which are always turned away from the Earth.

It is, I think, justifiable to say a little more about these 1959 vehicles, because they were so significant. Even after Sputnik 1 there were still people who, while conceding that artificial satellites could turn out to be very useful, treated the idea of a lunar landing with scorn, but Luna 1 made practically everyone realize that the Moon was accessible. The probe was launched on 2 January, and when it had reached a distance of 70,000 miles it sent out a cloud of sodium vapour which was duly photographed. The radio signals from the Luna were loud and clear, and were received not only in the USSR but also at Jodrell Bank, Britain's great radio astronomy observatory master-minded by Sir Bernard Lovell. Thirty-four hours after launch, Luna 1 bypassed the Moon and sent back data from relatively close range. Then it moved on, beginning a never-ending journey round the Sun. If the Soviet calculations are correct (and there is no reason to suppose otherwise), it has an orbital period of 446 days, and a path which may take it within six million miles of Mars. Of course, we have no hope of finding it again – signals from it were lost only 62 hours after the lunar rendezvous – but its task had been nobly done.

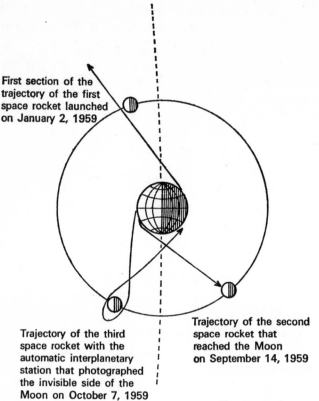

First section of the trajectory of the first space rocket launched on January 2, 1959

Trajectory of the third space rocket with the automatic interplanetary station that photographed the invisible side of the Moon on October 7, 1959

Trajectory of the second space rocket that reached the Moon on September 14, 1959

Fig. 35. Paths of the first Lunas

Luna 2 was a different kind of vehicle. It began its journey on 12 September, and, like its predecessor, it sent back information of various kinds during its flight; its programme included studies of cosmic rays, investigations of particles sent out by the Sun, possible magnetic fields in space, the numbers of meteoric particles, and methods of rocket control and guidance. The absence of any appreciable magnetism associated with the Moon was confirmed, which was only to be expected.

Before hitting the Moon, the last stage of the compound rocket was separated from the container carrying the scientific equipment. Presumably, no effort was made to bring either component down gently; this, remember, was in the Neolithic period of space research! The predicted time of impact was 2100 hours GMT on 13 September, and observers all over the world were at their telescopes, hoping to see some sign of a flash as Luna 2 reached the end of its journey. The results from the Jodrell Bank 250-foot radio telescope showed that as it neared its target the probe was still sending back loud, clear signals. At 21 hours 2 minutes 23 seconds the signals ceased abruptly. This, then, was the moment of impact.

No signals were received from Luna 2 after landing, and the exact impact site is not known, but it was somewhere near Archimedes in the Mare Imbrium. In the future, no doubt, someone will go there and collect the shattered remains of the first man-made vehicle to touch down on the Moon.*

Next, in the following month, came Luna 3, the first probe to send back really vital information about the Moon's surface. It had one main task: to go round the Moon, and tell us what the enigmatical far side was really like.

The question of 'What lies on the other side of the Moon?' had been tantalising astronomers for many centuries. It was so infuriating not to know; the Moon, after all, is extremely close to us, but until October 1959 our knowledge of its surface was restricted to the 59 per cent which can be studied from Earth. The remaining 41 per cent was absolutely inaccessible, and nobody knew what it was really like.

Speculation was rife, and some of the ideas put forward were fascinating. There was, for instance, a theory due to a famous last-century Danish mathematician, Hansen, who was busy studying the movements of the Moon when he found some discrepancies which he could not explain. They led him on to suggest that the Moon is not uniform in density, but has one hemisphere slightly more massive than the other. This lopsidedness would shift the centre of gravity some way from the centre of figure, and he worked out that it was 33 miles further from the Earth. He concluded that all the Moon's atmosphere and water had been drawn round to the far side, which might well be inhabited!

Even Sir John Herschel, the most famous astronomer of the time, was impressed, and in 1870 he wrote that on the Moon's far side there might be an ocean of ordinary water.

> The lunar atmosphere would rest upon the lunar ocean, and form in its basin a lake of air,** whose upper portions would be of excessive tenuity... It by no means follows then, from the absence of visible indications of water or air on this side of the Moon, that the other is equally destitute of them, and equally unfitted for maintaining animal or vegetable life.

---

* I must reluctantly gloss over other reports, such as one from an earnest lady who telephoned me saying that she had been watching the impact with binoculars, and had seen the Moon split in half. Incidentally, the generic term for the probes in common use at that time was 'Lunik'.

** Herschel's italics (in the original, but here shown as roman).

All this sounds rather strange in the light of what we now know, and Hansen's theory was soon abandoned. It is true that the centre of mass is slightly displaced from the centre of figure, but the amount is less than two miles. The situation was summed up rather neatly by a famous poem which was, I believe, written by a housemaid with literary aspirations, and which has been handed down to posterity. There seem to be several versions of it, and I have chosen the most quoted one:

> *O Moon, lovely Moon with the beautiful face,*
> *Careering throughout the bound'ries of space,*
> *Whenever I see you, I think in my mind*
> *Shall I ever, O ever, behold thy behind?*

All that could really be said was that the far side was likely to be just as barren, just as hostile and just as airless as the side which we have always known. When it came to the question of the arrangement of craters and mountains, or the frequency of maria, astronomers could do little more than make intelligent guesses.

During the immediate post-war years, various observers were doing their best to plot the libration regions near the very edge of the Moon's disk. It was not easy, because everything was horribly foreshortened, and it was difficult to tell a crater from a ridge; also, one had to take advantage of the rare occasions when the libration in any particular region was favourable at the time of suitable solar illumination. All the same, it was immensely fascinating, and there was always the chance of making a discovery. I did so myself on one occasion, when I happened upon a large walled plain right on the limb at maximum libration – beyond the huge, ruined formation Otto Struve (shown on Map 8 in Appendix 5). I duly reported it, and it was subsequently photographed; it has been named Einstein. I also found what appeared to be a small 'sea' right on the limb, and I suggested the name of Mare Orientale. What I did not know, of course, was that it is a vast ringed formation,

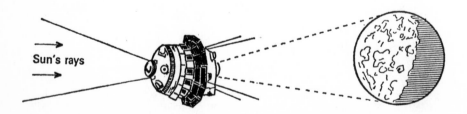

Fig. 36. Position of Luna 3 during the photography of the far side of the Moon

extending on to the Moon's far side; only a fraction of it can ever be seen from Earth. I was lucky enough to be plotting that part of the limb at the time of maximum favourable libration.

Like its predecessors, Luna 3 was given a full research programme, but its main task was to pass beyond the Moon and photograph the far side (fig. 36). The launcher was one of the conventional step-vehicles; it had to be powerful, because of the weight of its load. Without fuel, the upper stage of the rocket weighed about a ton and a half, while the weight of the 'station' itself amounted to nearly nine hundredweight – massive by 1959 standards.

All went well. By 4.30 GMT on 7 October the rocket had passed by the Moon, and lay beyond it, at a distance of under 40,000 miles from the lunar surface. The photographic apparatus was switched on, and for the next forty minutes the pictures were taken. Two cameras were used, giving photographs on different scales. After the programme had been completed, the films were automatically developed and fixed ready for transmission back to Earth.

Delay was inevitable, because Luna 3 was still receding from us. It reached its apogee or furthest point on 10 October, when it was 292,000 miles away, and then started to swing in once more, reaching perigee (29,000 miles) on 18 October. It was then that the pictures were sent back. They were scanned by a miniature television camera, and the transmissions were picked up by the waiting Russians. Late on 24 October, the photographs were given to the world.*

Blurred and lacking in detail though they are by modern standards, the Luna 3 pictures represented a tremendous technical triumph. Several features on the far side showed up, notably a dark-floored walled plain which the Russians christened Tsiolkovskii in honour of the famous rocket pioneer. Another dark feature was named the Mare Moscoviense or Moscow Sea. Bright craters were also detectable, with some ray-systems. Inevitably, there were errors in interpretation – a suspected high mountain-chain, which was named the Soviet Range, later turned out to be non-existent – but it was an encouraging start. Evidently the Russians hoped to rerun the pictures later on, but contact with Luna 3 was lost abruptly and was never regained, so that we do not know what happened to it.

The next four years were less fruitful. At this stage, the American lunar-probe programme was frankly floundering; the vehicles either went out of control, exploded, or missed the Moon completely. The Russians had a similar

---

* I shall not forget my first sight of them. At 10.15pm I was just starting a live broadcast in my BBC television series *The Sky at Night* and the pictures came through direct on the screen, giving me no time at all to think out a suitable commentary. Luckily the Mare Crisium was shown clearly, though naturally from an unfamiliar angle, and I was able to get my bearings.

failure with Luna 4 in April 1963; they may have been trying for a soft landing on the surface, but with no success. The next real achievement came with America's Ranger 7, which hit the Moon on 31 July 1964. Inevitably it destroyed itself, but during the last minutes of its flight it sent back over 4,000 high-quality photographs, and for the first time the surface of the Moon could be studied from really close range.

The impact point was in the Mare Nubium or Sea of Clouds, near the 36-mile, low-walled crater Guericke (Map 12 in Appendix 5). (The region was promptly named the Mare Cognitum or Known Sea, though it does not seem that this name has found its way on to many maps.) Four cameras were used altogether, and were in operation for only just over a quarter of an hour, so that the area covered photographically was somewhat restricted – about the same as that of France. The last picture was transmitted only 0.19 of a second before impact, and showed a region of the Moon measuring 105 feet by 150 feet, with crater-pits down to a few inches in diameter. There were many small features, and some rounded depressions, almost within walls, which looked like collapse features. Early in 1965 two more Rangers were equally successful. Number 8 (20 February) landed in the Mare Tranquillitatis, and No. 9 (24 March) inside the walled plain Alphonsus, which was – and is – of special interest because it is thought to be one of the more active areas of the Moon.

The results from these three crash-landers were much the same, so I can deal with them jointly. Even the smooth-looking parts of the Moon turned out to be anything but level when seen from point-blank range; there were pits, hummocks and ridges everywhere. Some of the rills inside Alphonsus proved to be made up of chains of small craters which had run together, and the whole lunar surface was rock-strewn. This brings me on to another controversy which was very much to the fore in the early 1960s. Would the Moon's crust be firm enough to support the weight of a space-craft?

I have already mentioned the 'deep dust' theory, proposed by Dr Thomas Gold, one of the world's leading astronomers. He went so far as to claim that a space-craft 'would simply sink into the dust with all its gear'. Naturally his opinions carried a great deal of weight, and space-planners were decidedly apprehensive. To achieve a successful touch-down, only to lose the vehicle in a matter of seconds, would be most disappointing, but if Gold's theory were correct nothing much could be done.

Practical observers were not impressed. According to Gold the lunar dust would flow 'downhill', coming to rest in the lowest-lying parts of the Moon; of course the maria are at lower levels than the bright areas. Yet there are some craters on the maria whose floors are of exactly the same texture as the surface outside. Archimedes, on the Mare Imbrium, is one. To invade the crater, Gold's dust would have had to have climbed over the walls; in other words it would have had to have flowed uphill, which did not seem very likely. Also, dust would have dropped into the rills and produced dark floors, whereas in fact rill-bottoms are bright. There were other objections too, and I for one had no faith whatsoever in the alleged dust-layer; but there was only one test – send a probe to find out.

This is precisely what the Russians did, in February 1966. After several failures they launched Luna 9, which was slowed down by rocket braking while still well above the Moon, and dropped gently on to the surface near the edge of the Oceanus Procellarum, not very far from the dark-floored walled plain Grimaldi. Within a few minutes the first signals were being sent back from the lunar surface, and were received both in the Soviet Union and by Sir Bernard Lovell's team at Jodrell Bank. One fact emerged immediately. Luna 9 was standing on a hard layer, with no tendency to sink, so that Gold's dust theory was completely wrong. The scene was remarkably like that of a lava-field in Iceland or some similar place; here were various rocks and boulders strewn around, and the whole landscape was rough. Potential astronauts felt comforted. If a manned craft were to land, the Moon would at least refrain from swallowing it up.

As has so often happened, a Russian success was quickly followed by something comparable from America, and between June 1966 and January 1968 seven Surveyors were sent to make soft landings on the Moon. Five were successful; of the others, No. 2 went out of control at the critical moment and crashed to destruction, while contact with No. 4 was lost before touch-down. During the same period the Soviet team despatched Luna 13 (December 1966), which came down safely in the Oceanus Procellarum.

The results from these various vehicles were all much the same. The firmness of the lunar ground was confirmed, and many excellent photographs were sent back; even the first Surveyor managed more than 11,000, and contact with it was kept up for seven months.

Surveyor 3, launched on 17 April 1967, was also aimed at the Oceanus Procellarum, some 230 miles south of Copernicus. Its programme was not confined to photography. It carried a sort of mechanical scoop, so that the texture and mechanical properties of the lunar 'soil' could be tested. I mention Surveyor 3 particularly since over two years later Charles Conrad and Alan Bean, from Apollo 12, landed so close to it that they were able to walk over and examine it; Surveyor was apparently undamaged, though of course, its power had long since failed, and the astronauts were able to hack pieces off to bring home for analysis.

The last three Surveyors were even more fruitful, since they carried what may be called chemical sets, and were able to confirm the long-held suspicion that the lunar surface is made up essentially of grey volcanic rock known to geologists as basalt. Numbers 5 and 6 landed in relatively smooth areas (the Mare Tranquillitatis and the Sinus Medii respectively), but Surveyor 7 touched down on the northern outer slopes of Tycho, so that it was the first probe to visit a highland area. Analyses showed that there was more aluminium but less iron than in the maria, though the composition of the material was still essentially basaltic.

Valuable though these soft-landers were, they were matched by the American Orbiters, of which there were five – all highly successful. The Orbiter vehicles were put into closed paths round the Moon, so that full photographic coverage could be achieved. So many pictures were obtained that even today many of them are stacked away to await analysis. Orbiter 1 was launched on 10 August 1966. The fifth and last probe of the series began its journey on 1 August 1967; and when it came to the end of its career the task of mapping the Moon, begun by Harriot and Galileo so long before, was to all intents and purposes complete. Practically the whole of the Moon was photographed; the only region left out was a small part of the south polar zone, which was, fortunately, covered by similar Russian orbiting vehicles. The far hemisphere was mapped as adequately as the region accessible from Earth, and this therefore seems the right moment to say a little more about the Moon's 'other side'.

The overall aspect is not the same as that of the familiar hemisphere. It has been said that the far side is all highland, which is a very fair description of it. Large maria are entirely absent, though it is true that the Mare Orientale extends well over the limit of the region observable from Earth, and the same is true of a few of the minor seas along the eastern limb of the disk as we see it (that is to say, beyond the Mare Crisium; see Map 1 in Appendix 5). The so-called Mare Moscoviense and the rather smaller Mare Ingenii, on the far side, are not genuine maria. On the other hand there are various large basins, with light-coloured floors not filled with mare material. It seems that the crust is thicker on the far side than on the Earth-turned side, which may explain why lunabase has not welled up through the crust to the same extent. The centre of gravity of the Moon really is displaced away from the geographical centre, and is further away from the Earth; but the difference is less than two miles, so that there is no suggestion of a return to Hansen's theory.

There are some splendid craters, plains and valleys on the far side. One particularly impressive crater-valley is associated with the large formation which has been named Schrödinger, only just out of view from Earth. As I

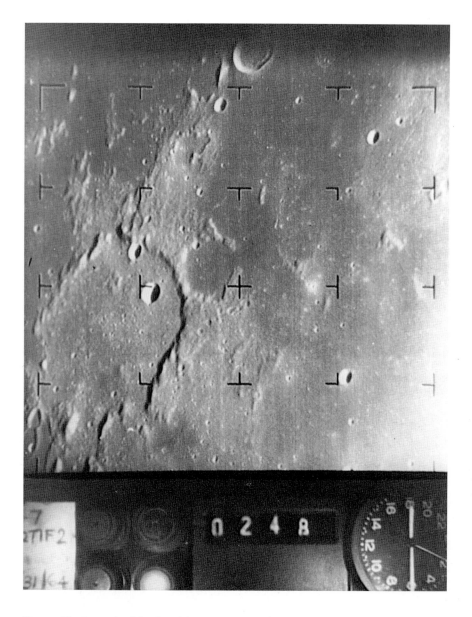

Above: Photograph of the Guericke region taken from
the Ranger spacecraft just before impact.

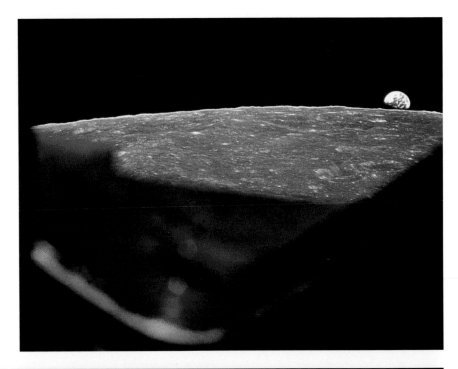

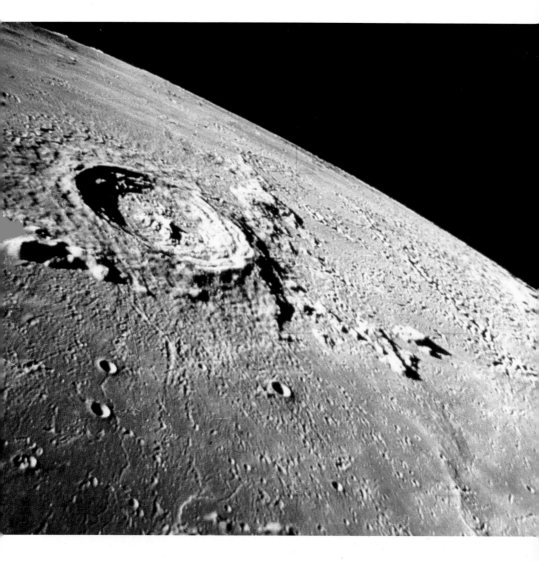

Opposite above:
An Apollo picture of
Earthshine over the Moon.

Opposite below:
Photograph of the Earth
over the Moon, sent back
from Apollo 8 in December
1968. The signature at the
bottom is that of Frank
Borman, Commander of
the mission.

Above: Eratosthenes –
one of the great lunar
craters – beautifully shot in
this Apollo photograph.
Note the terraced walls
and the central peak.

Above: An Apollo photograph of the far side of the Moon, with the Mare Orientale in the centre. The dark mass of the Oceanus Procellarum, visible from Earth, is to the lower left.

Oppsite: An oblique view of part of the Hyginus Rill, first discovered by Schröter. This photograph was taken by Apollo 10 astronauts.

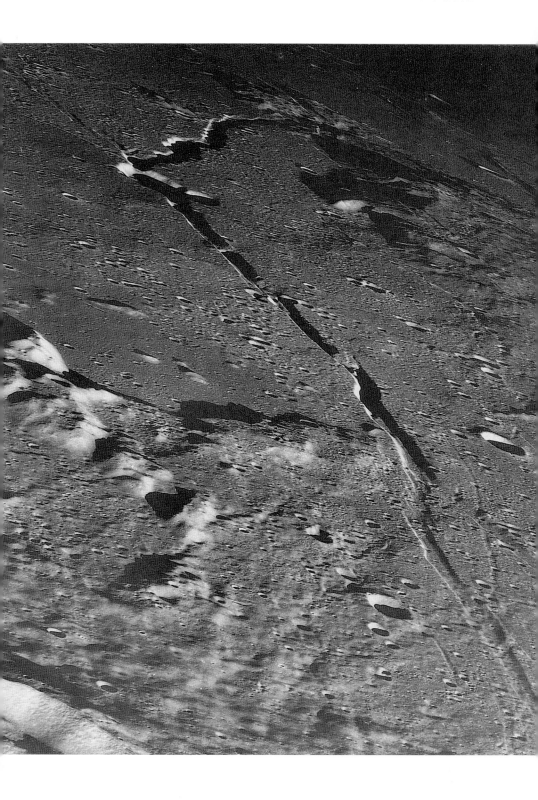

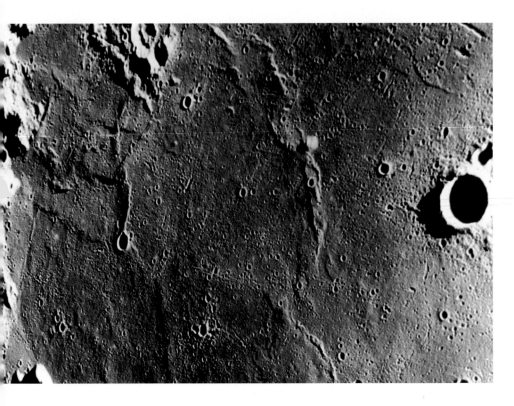

Above: Apollo landing site no. 3, Central Bay, with the crater Bruce at the right, seen from Apollo 10.

Opposite above: This is the crater Dædalus on the far side of the Moon, 50 miles in diameter, 179°E, 5.5°S, photograph taken from Apollo 11.

Opposite below: An oblique view of the lunar far side at 155°E, 10°S, photographed from Apollo 10.

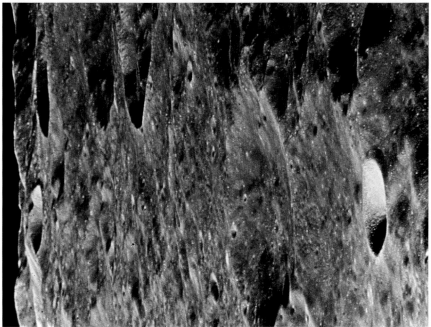

Above: The lunar module descending from Apollo 11, with Earth rising over the Moon.

Opposite: Buzz Aldrin standing on the Moon at the Sea of Tranquillity, in this famous photograph taken by Neil Armstrong, Apollo 11, July 1969. The scene is reflected in the transparent helmet.

Above: Apollo 15 astronaut James Irwin salutes the American flag against a typical lunar scene. The lunar rover is on the right of the lunar module, centre, Mount Hadley in the background.

Left: Neil Armstrong's footprint on the Moon: Apollo 11. The footprint will remain until covered by meteoric dust, unless it is taken away to a lunar museum.

Opposite: Close-up view of the little crater Linné, once suspected of showing signs of change. In fact, it is a normal, small impact crater. An Apollo 15 photograph.

Above: Apollo photograph of Aristarchus, the most brilliant crater on the entire Moon, 23 miles in diameter.

Opposite above: The lunar rover of Apollo 15, in the foothills of the lunar Apennines.

Opposite below: Orange soil, seen by Apollo 17. It does not indicate recent volcanic activity, but is due to very large numbers of very ancient glass particles.

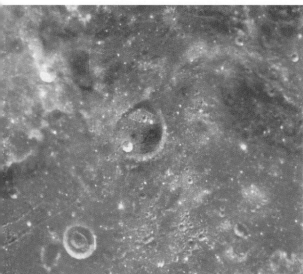

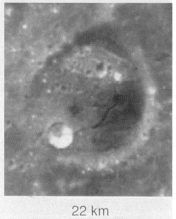

Apollo Basin Interior

22 km

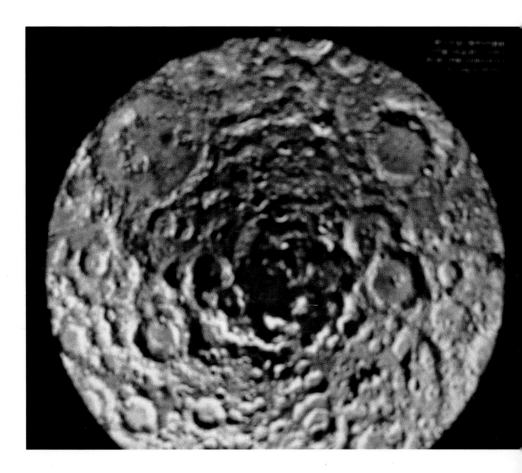

Opposite above: The Apollo 17 lunar rover vehicle at work, Mount Hadley in the background.

Opposite below: The Apollo Basin photographed from Clementine; the close up inset shows the scale of the photograph.

Above: The lunar north pole, photographed from Clementine. Some of these crater floors are permanently in shadow.

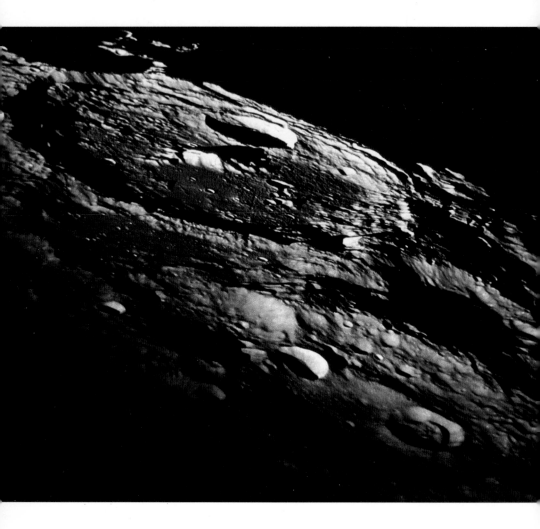

Above: A large crater on the far side of
the Moon, photographed from Apollo 13.

predicted long ago (probably more by luck than by judgement) the arrangement of the craters is less regular, though the general laws of distribution still apply; it is usually a smaller crater which breaks into a larger, not vice versa.

Tsiolkovskii is probably the most imposing object on the far side, if we except the Mare Orientale. The main characteristic of Tsiolkovskii is the darkness of its floor; in many photographs it gives the impression of being shadow-filled, though the real cause of the darkness is purely the colour of the interior itself. Tsiolkovskii breaks into another larger, less regular basin, Fermi, whose floor is not of the usual light hue. Note also the vast South Pole-Aitken basin, which is over 1,500 miles across. It was recorded by the Orbiters, though more detailed surveys were made in 1994 by the Clementine probe – about which I will have more to say later.

One minor controversy has centred around the naming of the various features on the far side. Since the Russians were first in the field, with Luna 3, they felt entitled to allot what names they liked to the objects they managed to identify, and nobody was in a position to argue. Most of these names are still recognized. Later the Americans joined in, and the whole matter was handed over to the International Astronomical Union, the controlling body of world astronomy. The system of naming lunar features after famous personalities has been followed, and is satisfactory enough.

Quite apart from the photographic results, the Orbiters (and the later Apollos) sent back data of all kinds including measurements of the shape of the Moon. They have found that the Moon is slightly egg-shaped; the axis pointing to Earth is between 1 and 2 miles longer than the other axis. The movements of the probes also led to the discovery of the mascons, which I mentioned earlier.

The first announcement of mascons came in 1968, with some work by the American astronomers P. Muller and W. L. Sjögren, who were studying the movements of Orbiter 5. They kept a careful check on Orbiter during eighty consecutive revolutions round the Moon, each taking 3 hours 11 minutes, and found that the velocity in orbit was not constant. There were times when there was a slight speeding-up, followed by a slowing-down to the original rate. The effect was very small indeed, but it could be measured.

If a probe passes over a region where the surface material of the Moon is unusually dense, then the probe will be pulled along, so to speak, and it will speed up. Muller and Sjögren found that this happened over the same areas during each revolution round the Moon, and this led them to assume the existence of dense masses below the lunar crust. It was soon found that there was strong correlation with the lunar seas, which are depressed below the upland level – the Mare Crisium and the Mare Smythii, for instance, by as much as two and a half miles. The term 'mascon' soon came into general use.

Mascons have been found below the regular seas (Imbrium, Crisium, Smythii, Serenitatis, Humorum, Nectaris, Humboldtianum, Orientale and Sinus Æstuum) and also below a few dark-floored walled plains, of which Grimaldi is probably the best example. It seems, then, that the extra mass (producing what is termed a positive gravity anomaly – that is to say, a greater pull than average) is associated with the material which filled the mare basins more than 3,000 million years ago. It also supports the idea that features such as Grimaldi are of the same type as the maria. On the other hand, craters of the Copernicus type tend to have negative gravity anomalies, and the same is true of the basins, which are not filled with mare material. Most of these unfilled basins lie on the far side of the Moon.

From what I have said here, it is obvious that our knowledge of the Moon was increased tremendously between 1959 when Luna 1 made its flight, and 1967–8, when the Orbiter series came to an end. But there is a vital difference between exploring with automatic probes, and going to see for oneself; and even while Orbiter 5 was still transmitting, the programme of manned travel was well under way. America had announced her intentions, and had started to put them into practice. The real attack on the Moon was about to begin.

15

Apollo

On 20 July 1969 – how long ago that seems now! – Neil Armstrong stepped out on to the surface of the Moon. It was a moment never to be forgotten. Remember, there was no provision for rescue; if the lunar module had a faulty landing, or if it had come down at an angle which made it impossible to take off again, the results for the astronauts would have been dire. As I have said, I was in the main BBC Television studio, carrying out a live commentary; when I heard Neil's voice – 'The *Eagle* has landed' – my feeling was one of overwhelming relief. The dangers were very far from over, but at least one major obstacle had been overcome.

The Apollo programme had begun much earlier, with President Kennedy's declaration that a man could be sent to the Moon before 1970. The first true Apollo (No. 7; the earlier six had been unmanned tests) orbited the Earth in October 1968. In the following December Apollo 8, carrying Astronauts Borman, Lovell and Anders, went right round the Moon, and made the first direct observations of the far side. Apollo 9 was another Earth Orbiter; Apollo 10 was the final test, and Astronauts Stafford, Young and Cernan carried out the whole programme apart from the actual landing.* All was ready for Apollo 11. Neil Armstrong and Buzz Aldrin were scheduled to come down on the surface, while Michael Collins would continue orbiting the Moon in the third stage of the space-craft.

Sadly, we have to admit that the support given to Apollo by the US Government was not entirely altruistic. The Russians had their own manned lunar programme, but their rockets proved to be too unreliable, and after several disastrous mishaps their 'N' programme was abandoned. Their programme has since developed along different lines. Instead of sending men to the Moon, they have sent soft-landers, recoverable probes, and 'crawlers'. To date, two there-and-back vehicles have been sent successfully to the surface – Luna 16 in 1970 and Luna 20 in 1972; both came down in the

---

* Again I was carrying out a live television commentary, and I remember my actual words at the vital moment. 'The men of Apollo 8 are now on the far side of the Moon; they are completely out of touch, and they have carried out a dangerous manœuvre. As soon as they come round the Moon's limb, we should hear their voices. I will say no more now; in less than thirty seconds we should pick up the messages from the first men round the Moon. This is one of the great moments in human history.' And with an exquisite sense of timing, the BBC producer switched over to the children's programme *Jackanory*.

eastern hemisphere, in the general area of the Mare Fœcunditatis and the highlands of Apollonius, and both brought back rock samples, though the quantities were naturally very limited compared with the Apollo hauls.

The crawlers, taken to the Moon in Luna vehicles, were in a different category. Lunokhod 1, looking distinctly like an antique taxicab, landed in the western part of the Mare Imbrium in November 1970, and for months it moved around the surface, guided by its controllers in the USSR and sending back valuable information as well as a great many pictures; finally its power gave out, but no doubt it will be collected one day, because we know exactly where it is. Lunokhod 2 had as its target the semi-ruined crater-bay Le Monnier, on the eastern edge of the Mare Serenitatis; altogether it covered a traverse length of almost 20 miles. The region surveyed was only 110 miles north of the site of the Apollo 17 landing, which had taken place a month earlier. Weird though they looked, the Lunokhods were brilliantly successful, and it is rather surprising that no more were launched before the end of the century.

So far as Apollo was concerned, almost everything depended upon the results of that first trip. The two astronauts who actually landed showed themselves to be expert scientists as well as explorers, and Neil Armstrong was right when he spoke those immortal words: 'That's one small step for a man, one giant leap for mankind.' During their one foray outside the module, the astronauts set up an ALSEP (Apollo Lunar Surface Experimental Package) which contained instruments of various kinds and which operated for some time after the end of the mission.

I doubt if anyone has bettered the description given by the second Apollo astronaut, Buzz* Aldrin, soon after he had followed Neil Armstrong down to the surface: 'Magnificent desolation.' Neil's words were equally memorable:

> You generally have the impression of being on a desert-like surface, with rather light-coloured hues. Yet when you look at the material from close range, as in your hand, you find that it's really a charcoal grey. We had difficulties in perception of distance. For example, from the cockpit of the lunar module we judged our television camera to be only 50 or 60 feet away, yet we knew that we had pulled it out to the full extent of a 100-foot cable. Similarly, we had difficulty in guessing how far the hills on the horizon might be away from us. The peculiar phenomenon is the closeness of the horizon, due to the greater curvature of the Moon's surface — four times greater than on Earth; also it's an irregular surface, with crater rims overlying other crater rims.

---

* This is his correct name. He was christened Edwin, but changed officially to Buzz after the Apollo mission.

There were no problems on the surface; moving about under one-sixth Earth gravity was easy enough, though everything seemed to happen in slow motion. Neither were communications difficult; the entire trip was shown on television – even the conversation between President Nixon and the astronauts on the Moon, a quarter of a million miles away.

As expected, the rocks brought back were composed of familiar substances, and were essentially basaltic. Before being exposed to the air they were very carefully examined, just to make sure that they contained nothing harmful, and the astronauts themselves were quarantined until the experts were satisfied. The chances of bringing back anything dangerous were very slight, but they were not nil – remember Professor Quatermass! Quarantining was abandoned after Apollos 11 and 12, because it was clear that the Moon is, and always has been, completely sterile. The situation will be different when we bring back samples from Mars, but this is unlikely to be before 2004 at the earliest, so that we have plenty of time to make plans.

Apollo 12 followed in December 1969, and was equally successful; the landing site was in the Oceanus Procellarum, and Astronauts Conrad and Bean were able to walk over to the old probe Surveyor 3, which had come down there in 1967, and bring parts back for analysis. Apollo 13, sent up in April 1970, was a near-disaster. During the outward journey there was an explosion in the service module of the space-craft, which blew a gaping hole in the side of the vehicle and put the main engines permanently out of action. The lunar landing was promptly called off, and it was only by brilliant improvisation, both by the technicians at Mission Control and by the astronauts themselves, that the crew were able to come home safely; it was a timely reminder that space is a dangerous environment. Disaster would have been inevitable if the explosion had happened on the way back from the Moon, because the lunar module would have been jettisoned. In the event, it was the engine of the lunar module which saved the situation.

There were four more missions. Apollo 14 (January/February 1971) was commanded by Alan Shepard, who had been the first American in space only ten years earlier; he and his companion, Edgar Mitchell, spent more than nine hours outside the lunar module, and used a 'cart' to carry their equipment across the landscape near the ruined crater Fra Mauro. Apollo 15 (July 1971) took David Scott and Jim Irwin to the foothills of the Apennines, near the spectacular Hadley Rill, and this time they drove around the surface in a 'moon car' or Lunar Rover. The Rill is over 60 miles long, almost a mile wide and a third of a mile deep; there were obvious signs of stratification, giving the impression that it has been cut by a lava-stream during the Moon's active period. The impressive peak of Hadley Delta was in full view, and proved to be quite unlike the rough, jagged mountains described by science-fiction writers in the pre-Space Age days; as I have noted, Scott even called it 'a featureless mountain'. The photographs and television pictures make it seem much closer to the Rover than it actually was; the distance to the base of the peak was over 20 miles. As all the astronauts have found, distances on the Moon are hard to estimate, partly because the lack of atmosphere means that there can be no softening of shadows, and partly because the Moon is smaller than the Earth, so that its surface curves more sharply and the horizon is correspondingly closer. Incidentally, many people have asked why the pictures show no stars in the black sky. As Neil Armstrong told me, the reason is glare from the surface rocks. Shield your eyes and dark-adapt for a few minutes, and the stars come into view.

Apollo 16 (April 1972) landed in the highlands of Descartes, and the Moon-walkers, John Young and Charles Duke, undertook an elaborate scientific programme. Finally, in December 1972, came Apollo 17, which touched down in the region of Taurus-Littrow. The mission commander, Eugene Cernan, had already been round the Moon, in Apollo 10; his companion, Dr Harrison Schmitt, was a professional geologist who had been trained as an astronaut specially for the occasion. Predictably, his expertise proved to be invaluable.

Among the features found in the lunar materials are vast numbers of small glassy particles, or 'beads' of microscopic size. Their colours are varied, and so are their forms. They seem to have been produced by what is termed shock metamorphism; a sudden, violent disturbance in the material can cause alterations in structure, plus melting, and glass is the result. Meteorite impacts could certainly account for this.

I was in Mission Control, Houston, during the Apollo 17 mission. On 12 December Cernan and Schmitt were driving in their Lunar Rover out to South Massif, one of the two mountains overlooking the valley, where the material was made up chiefly of breccias, with sub-floor rocks of basaltic nature. On the way back their route took them to a small crater which had been given the unofficial name of Shorty, and suddenly, we heard Schmitt's voice: 'It's orange – crazy!' There was a band of orange-coloured material circumferential to the crater, and Schmitt compared it with a fumarole effect, which would indicate relatively recent volcanic activity. Samples were collected, but when they were brought home and analysed the explanation proved to be very different. The orange colour was due to tiny coloured glassy beads, which were very old (about 3,800 million years), and fumarole activity was not involved. Nothing like it had been found elsewhere; the colour was so pronounced that it was easily visible on our television screens at Houston when Schmitt was examining it, a quarter of a million miles away from us.

Altogether Cernan and Schmitt spent over twenty hours driving around. The Lunar Rover performed well, but surface dust was a nuisance. To quote Gene Cernan: 'the dust is like graphite, but graphite lubricates, while lunar dust makes things stick together. It gets into your space-suits and all moving parts of your vehicles. The dust is so fine that it even gets into the pores of your skin. It took me many weeks after my return to get rid of the last traces of it.' But at least the dusty layer was very shallow indeed, and no drifts were ever found.

When Cernan and Schmitt blasted away from the Moon, to rejoin Ron Evans in the orbiting section of the space-craft, the first stage of Man's exploration of the Moon was over. Three more Apollos had been planned, but were never sent, partly because of financial cutbacks but mainly because Apollo had done almost everything that it was capable of doing. Moreover, there was always the fear that something would go badly wrong. There must surely be provision for rescue before any more astronauts undertake the journey.

Obviously, the Apollo programme has increased our knowledge of the Moon beyond all recognition. For example, we now know what the surface materials are like. All the rocks brought home for analysis are igneous, or breccias produced by impact processes; there are no sedimentary or metamorphic rocks. In the lavas, basalts are dominant, containing more titanium than terrestrial lavas; small amounts of metallic iron were found. Many lunar rocks are also rather poor in sodium and potassium, though there is a different type of basalt called KREEP, which contains potassium (chemical symbol K), Rare Earth Elements, and phosphorus. A new mineral – an opaque oxide of iron, titanium and magnesium – has been named armalcolite in honour of the three Apollo 11 astronauts Armstrong, Aldrin and Collins. The highland rocks are between 4,000 and 4,200 million years old; over 90 per cent of the surface dates back for over 3,000 million years.

Quite apart from studies of the Moon itself, there were unrivalled opportunities for carrying out researches into other branches of science. The lack of atmosphere, the absence of any appreciable magnetic field, and the low gravity combine to make the Moon an ideal site. For example, consider the solar wind, which is made up of a plasma (that is to say, ionized gas) streaming out from the Sun in all directions. It hits the lunar surface, unchecked by the atmospheric and magnetic barriers which shield the Earth, and darkens the outer materials; it can be collected by suitable equipment, and samples of it have been brought back for analysis. It has even been said that 'the Moon collects the Sun for us' – because the solar wind is a sample of the Sun. Another interesting experiment was designed to detect gravitational waves coming from deep space, which may or may not exist; at present the evidence is very inconclusive.

When we return to the Moon, we will find that the equipment left by Apollo is there waiting for us; there is nothing to damage it. No doubt the Lunar Rovers will be given new batteries and driven happily off to the Base which will presumably be set up. The same applies to the Russian Lunokhods, and at least we may be sure that the next expeditions will be truly international. The men of Apollo have shown the way, and they will never be forgotten.

16

The Search for Ice

It would not be quite correct to say that the mapping of the Moon was complete after the Orbiter and Apollo missions. In particular, the polar zones were not covered as well as the rest of the surface, and there were also extra problems. Some of the polar craters are very deep by lunar standards, and this means that their floors are always in shadow. Obviously no details can be seen inside them, and they remain bitterly cold; everything is inky black. The temperatures may be as low as -230 degrees C, only a few tens of degrees above absolute zero, making the floors of the polar craters among the chilliest places in the entire Solar System.

On 2 January 1994 NASA launched a new probe, Clementine, not from Canaveral but from the Vandenberg Air Force Base. (The name comes from an old mining song: Clementine is 'lost and gone forever'.) Vandenberg was chosen in preference to Canaveral because the mission was a joint venture between NASA and the US Strategic Defense Initiative Organization. The Pentagon wanted to test some military equipment under space conditions, but had to get round a United Nations directive which frowned upon anything of the sort. Taking the Moon into the programme was the answer, and Clementine was given a rather complicated programme. It would be put into lunar orbit on 24 February and would begin a comprehensive programme of mapping, with special emphasis on the polar regions, as well as analysing the surface materials at various wavelengths ranging from ultra-violet to infra-red. After two months it would blast away from the Moon and rendezvous with a small asteroid, Geographos, which is the most elongated natural body known; it is 3.2 miles long but only just over one mile broad. Geographos is classed as a PHA (Potentially Hazardous Asteroid) because even in its present path it can come within four million miles of the Earth, and the orbits of tiny bodies of this kind are always erratic. If Geographos did hit the Earth, it would do an immense amount of damage.

Things did not go completely according to plan. Clementine reached lunar orbit successfully, and continued mapping until 23 April; at one point it was only 260 miles above the lunar surface. Unfortunately a fault in the propulsion unit meant that the Geographos rendezvous had to be abandoned, but so far as the Moon was concerned, Clementine worked excellently, and the images sent back were the best ever obtained. For example, a splendid view showed up the vast South Pole-Aitken Basin, which is the largest and deepest on the Moon – over 1,500 miles across, and 8 miles deep; it actually contains the lunar south pole. But there was one very unexpected development. On 3 December 1996 the Pentagon announced that ice had been found in the bottoms of some of the polar craters.

This would be a major discovery, and it caused great excitement, not only among NASA scientists but also among non-astronomers. It was claimed that if all the ice were melted it would fill a lake two miles square and 35 feet deep; a cubic yard of polar material could yield several gallons of water simply by heating it and evaporating it off. If the ice totalled more than 50 million tons it 'could still sustain several thousands of people for more than a hundred years, using the water in the same way as we do on Earth,' commented one NASA spokesman. Future Moon colonists would find it very useful indeed.

After Clementine came Prospector, launched from Canaveral on 6 January 1998 (this time there was no military involvement). Prospector was designed to continue and improve upon the Clementine results, and was very successful; the images were even better than Clementine's, and indications of ice were found at both the lunar poles. On 5 March there was a clear statement from Dr Alan Binder, Prospector's Mission Scientist: 'We have found water.' It was then estimated that between 2,600 million and 80,000 million gallons of frozen water existed. According to another NASA scientist, William Feldman, 'There are a bunch of craters filled up; with water ice. This is a significant resource that will allow a modest amount of colonization for many years. Water can now be mined directly on the Moon, instead of having to be shipped from Earth.'

It all sounded very promising, but almost at once uneasy doubts started to creep in. One sceptic was Dr Harrison Schmitt, the only geologist who has been to the Moon. My own view was quite definite. I did not believe in the presence of lunar ice even then, and I do not believe in it now.

First, bear in mind that Clementine and Prospector certainly had not detected an ice-sheet. The ice would be contained in the lunar material in thin concentrations – probably less than 1 per cent – over a large area, up to 18,000 square miles around the north pole and up to 7,200 square miles around the south pole. Mining it, particularly under those intensely cold conditions, would be far from easy. In any case, was it certain that ice existed at all? All that the probes had done was to detect hydrogen, which is a very different matter.

The instruments used were known as neutron spectrometers. When cosmic ray particles bombard the Moon – as they do, all the time – neutrons are produced, moving at very high velocities. However, if a neutron happens to hit a hydrogen atom it is slowed down, because it loses much of its momentum – just as a snooker ball will do after cannoning off another ball. Collisions with heavier atoms do not have the same effect, so if we find a great number of 'slow neutrons' we can assume that hydrogen is present. As Clementine and Prospector passed over the polar craters, they recorded many more slow neutrons than they would otherwise have done. It was then claimed that the hydrogen had combined with oxygen to produce water; other hydrogen compounds, such as methane, are much less stable, and would not persist.

Yet solar wind also bombards the Moon, and this contains hydrogen. If the spectrometers were recording solar wind, there would be no need for water, and the claim that ice had been definitely found did seem very premature. Moreover, how could the ice have found its way into the polar craters? It certainly is not 'lunar'. Samples brought back by Apollo astronauts and the Russian unmanned probes show absolutely no trace of any hydrated materials.

It is also worth mentioning the meteorites, found on Earth, which are widely believed to have come from the Moon, blasted away from the lunar surface by a violent impact long ago. The first of these was found in Antarctica, near the Allan Hills, and is known as ALH 81005 (Antarctica is an excellent site for meteorite-hunters because it has been left undisturbed until very recent times). ALH 81005 is an anorthosite breccia with a brownish fusion crust. Its make-up is very lunar in type, and the evidence that it does come from the Moon is strong, if not conclusive. More than a dozen similar meteorites have been found, not all of them in the Antarctic. None shows any trace of watery material.

The only alternative is to assume that the ice was deposited on the Moon by a comet. True, comets are icy bodies, but there are problems here too. The heat generated during the impact would probably have been enough to blast the material away from the Moon altogether, and it seems strange that any ice would be obliging enough to come to rest inside a polar crater.

Eventually NASA decided to carry out a drastic but fascinating experiment. At the end of its useful life Prospector would be deliberately crashed into a polar crater, in the hope that signs of water would be found in the column of débris thrown up; also, there could be signs of a nearly transparent haze of water vapour and hydroxyl which might drift above the lunar limb for hours after the dusty ejecta from the impact had fallen back to the surface. And so at 0952 GMT on 31 July 1999, Prospector met its fiery fate. Probably it did come down in the area selected, but telescopes showed no débris, and Earth-based spectrometers were no more successful. In fact, the results of the experiment were entirely negative.

To my mind, there is one other objection to the ice theory which is fatal. The same claim had been made for polar craters on the planet Mercury, from surveys carried out by the Mariner 10 probes in 1974 and 1975. But identical results had been obtained from regions on Mercury which are not in permanent darkness, and where ice could not possibly survive even for a few seconds.

There, for the moment, the matter rests. In recent years there have been several lunar fly-by encounters by probes on their way to different targets, and there has even been one small Japanese impactor, Hiten, which hit the Moon in 1993, but nothing at all detailed. No doubt new space-craft will be launched in the near future, and the search will go on – but if any ice is found, I for one will be very much surprised.

17

Life on the Moon?

Can there be any natural life on the Moon? There is no air (or virtually none), the temperatures are extreme, and in every way the Moon is hostile. Yet remember that the first two Apollo crews were quarantined on their return, just in case they had brought back anything harmful. We might have been 99.99 per cent sure that there was no danger, but only after men had actually been there could we be quite certain.

Before going any further, it is only sensible to define just what we mean by 'life'. All life of the kind we can understand is based upon one element – carbon – and the same is probably true of life anywhere in the universe. Science fiction writers have taken great delight in describing creatures of entirely alien pattern – made up of gold, perhaps, and breathing pure hydrogen. Beings of this kind are generally referred to as BEMs (Bug-Eyed Monsters). Let us admit that we cannot rule out the possibility of BEMs. Shakespeare's line 'There are more things in Heaven and earth, Horatio...' holds good in science, as in everything else. A colleague of mine once commented that he could not deny the possibility of an intelligent extra-terrestrial being who looked like a cabbage and squeaked like a mouse. He did not think it at all likely, but it was not totally out of the question – and of course he was right. Yet when we start to consider these alien life-forms, speculation becomes both endless and pointless. If BEMs exist, then all our modern science is wrong; and this I am not prepared to believe when there is not a shred of evidence in favour of it. So for the moment I propose to discount all BEMs, and deal only with life as we know it.

Various conditions have to be met. There must be a reasonably even temperature, an atmosphere containing oxygen, and a supply of water, and none of these conditions are to be found on the Moon. All this seems to be so obvious today that we tend to forget that less than two centuries ago leading astronomers were quite ready to believe in a Moon capable of supporting intelligent beings.

The idea of a populated Moon is very old, and once it was realized that the Moon is a rocky globe it was tacitly assumed to be inhabited. Even the invention of the telescope did not bring about a prompt change of view. In 1634 came the posthumous publication of the famous *Somnium* or 'Dream', written by no less a person than Johannes Kepler (the man who first proved that the Earth moves round the Sun in an elliptical path), in which various weird and wonderful life-forms were described. Quite probably Kepler was deliberately letting his imagination run riot, but he and others of the period were very sympathetic to the idea that the Moon might have extensive oceans and a reasonably dense atmosphere. By 1800 the oceans had been abandoned, but it was still thought that there might be air and water, making surface life possible. Johann Schröter certainly thought so, and he was supported by the most famous astronomer of the day, William Herschel.

Herschel, the Hanoverian musician who became official astronomer to King George III, is best remembered for his discovery of the planet Uranus, in 1781, but his most important contributions were in the field of stellar astronomy (he was one of the first men to give a reasonable picture of the shape of our star-system or Galaxy). As an observer it is possible that he has never been equalled, and between 1781 and his death, in 1822, every honour that the scientific world could bestow came his way. His views about life in the Solar System were, then, rather surprising. He thought it possible that there was a region below the Sun's fiery surface where men might live, and he regarded the existence of life on the Moon as 'an absolute certainty'. In 1780 he wrote to the then Astronomer Royal, Nevil Maskelyne, as follows:

> *Perhaps conclusions from the analogy of things may be exceedingly different from the truth; but seeing that our Earth is inhabited, and comparing the Moon with this planet; finding that in such a satellite there is a provision of light and heat; also in all appearance, a soil proper for habitation fully as good as ours, if not perhaps better – who can say that it is not extremely probable, nay beyond doubt, that there must be inhabitants on the Moon of some kind or other?*

Maskelyne was not in the least convinced, and it is on record that later, when Herschel said much the same thing in a paper about lunar mountains, the Astronomer Royal deleted the offending paragraph before passing the paper for publication. Yet the idea of Moon-men was certainly not dismissed out of hand.

Schröter's views were not so extreme, but he too was reasonably sure that the Moon must be populated. He knew that the lunar atmosphere is thin, but he grossly overestimated its density, and even considered that some of the features visible on the Moon might be artificial. This last idea was supported by another German astronomer, Franz von Paula Gruithuisen (originator of the impact theory of crater formation), who announced in 1822 that he had discovered a real 'lunar city' on the borders of the Sinus Medii, not far from the centre of the disk.

Gruithuisen was a keen-eyed observer who did much excellent work, but unfortunately his vivid imagination tended to bring discredit upon him even in his own lifetime. His 'lunar city' was a case in point. He described it as 'a collection of dark gigantic ramparts... extending about 23 miles either way, and arranged on either side of a principal rampart down the centre... a work of art'. Actually, his 'dark gigantic ramparts' turn out to be no more than low, haphazard ridges. Two of them are vaguely parallel for some distance, but there is nothing in the least like an artificial structure, and in any case the ridges are so low that they are difficult to see at all except when near the terminator. There can be no question of any change here, as Schröter, years before Gruithuisen, and Mädler, ten years afterwards, drew the region just as it is today.

Beer and Mädler showed that the Moon is definitely unable to support higher life-forms, and after the publication of their great book, in 1838, the Moon-men were more or less handed over to the storytellers, who used them to the full. Bug-eyed monsters, however, are relative newcomers to the literary scene (H. G. Wells was mainly responsible for them), and up to the time of Herschel it was thought more likely that the 'Selenites' were humanoid.

Even so, the people of 140 years ago were quite ready to believe in bizarre life-forms. This led to the famous Lunar Hoax, the biggest scientific practical joke of all time, apart possibly from the Piltdown Man.

Sir William Herschel had explored the northern skies with his great telescopes, discovering vast numbers of double stars, clusters and nebulæ, and probing the depths of space as no man before him had ever done. However, the southernmost stars, which never rise over England, remained comparatively unknown. Catalogues of the brighter ones were drawn up from time to time, but by the nineteenth century it had become clear that there was an urgent need for a more detailed survey. Fittingly enough the task was undertaken by William Herschel's son, John.

On 13 November 1833, Sir John Herschel set out for the Cape of Good Hope, taking with him a large telescope (you can see it today at Flamsteed House, the old Greenwich Observatory). Herschel stayed at the Cape for four years, and when he finally left, in 1838, his work had been well done. It took him more than a decade to collect and sort all the observations.

Herschel did not mean to pay any particular attention to the Moon or planets, which can be seen just as well from the northern hemisphere as from the Cape. He was concerned with the stellar heavens, and there was more than enough to occupy him. However, Richard Locke, a graceless reporter of the New York *Sun*, had a bright idea. Herschel was on the other side of the world; communications in those days were slow and uncertain; who was there to check any statements that he might care to make?

Locke saw his chance, and took it. On 25 August 1835, the *Sun* came out with a headline reading 'Great Astronomical Discoveries Lately Made by Sir John Herschel at the Cape of Good Hope', and an account of how Herschel had built a new telescope powerful enough to show the Moon in amazing detail. Locke's article was so cleverly worded that it sounded almost plausible. It was well known, he wrote, that the chief limitation of any telescope is that it cannot collect enough light for extreme magnification, but Herschel had overcome this by effecting 'a transfusion of artificial light through the focal point of vision' – in other words, by using the telescope to form an image, and then reinforcing the image by means of a light-source in the observatory itself!

The way was open, and the *Sun* kept up the good work for the next six days. The lunar scenery was varied and colourful: 'A lofty chain of obelisk-shaped or very slender pyramids, standing in irregular groups, each composed of about thirty to forty spires, every one of which was perfectly square... they were of a faint lilac hue, and were very resplendent.' Next came animals, one of which was 'of a bluish lead colour, about the size of a goat, with a head and a beard like him, and a single horn... In elegance of symmetry it rivalled the antelope, and like him it seemed an agile sprightly creature running at great speed, and springing up from the green turf with all the unaccountable antics of a young lamb or kitten. This beautiful creature afforded us the most exquisite amusement.' On the next night the telescope was turned to 'a large branching river, abounding with lovely islands, and water-birds of numerous kinds... Near the upper extremity of one of these islands we obtained a glimpse of a strange amphibious creature of spherical form, which rolled with great velocity across the pebbly beach.' Botany was not neglected; in the crater Endymion 'Dr Herschel has classified not less than 38 species of forest trees', but in the crater Cleomedes there seemed to be no living creature, except for 'a large white bird resembling a stork'. There were 'hills pinnacled with tall quartz crystals, of so rich a yellow and orange hue that we at first supposed them to be pointed flames of fire... a pure quartz rock, about three miles in circumference, towering in naked majesty from the blue deep; it glowed in the sun almost like a sapphire'. The brilliant Aristarchus was 'a volcanic crater, awfully rivalling our Mounts Etna and Vesuvius in the most terrible epochs of their reign... we could easily mark its illumination of the water over a circuit of sixty miles'.

The climax was reached on 28 August, with Locke's priceless account of lunar bat-men:

Certainly they were like human beings... Having observed them for
some minutes we introduced lens Hz, which brought them to an
apparent proximity of eighty yards. They averaged four feet in height,
were covered, except on the face, with short and glossy copper-coloured
hair, lying snugly upon their backs. The face, which was of a yellowish
flesh-colour, was a slight improvement upon that of the large orang-
outang... The mouth, however, was very prominent, although
somewhat relieved by a thick beard upon the lower jaw. These creatures
were evidently engaged in conversation; their gesticulation, particularly
the varied action of their hands and arms, appeared impassioned and
emphatic. We hence inferred that they were rational beings.

Locke was clever enough to bring in 'science' now and then. Sometimes a higher-powered eyepiece had to be used, sometimes conditions were unsuitable for observing, and sometimes it was necessary to turn up the hydro-oxygen burners to light up the faint image by the method of 'artificial transfusion'. On one occasion the astronomers forgot to cover up the main lens, so that when the Sun shone on it the lens acted as a vast burning-glass and set fire to the observatory.

The articles met with a mixed reception, but some eminent critics swallowed the bait hook, line and sinker. 'These new discoveries are both probable and plausible,' declared the *New York Times*, while the *New Yorker* thought that the observations 'had created a new era in astronomy and science generally'. A women's club in Massachusetts is said to have written to Herschel asking for his views on how to get in touch with the bat-men and convert them to Christianity, while even the Academy of Sciences in Paris held a debate when the news spread across to Europe, though it must be added that the French astronomers were highly suspicious.

The hoax was exposed by a rival paper within a few days, and the *Sun* itself confessed on 16 September. Even then, however, lingering doubts remained, and not for some months was the whole absurd business finally killed. Herschel apparently took the joke in good part when he heard about it, which was not until some time later.

It is easy to laugh, but can we afford to? Remember the 1938 panic, when a misleading broadcast of H. G. Wells' *War of the Worlds* led some people in the United States to believe that the Earth was being attacked by monsters from Mars; and even more recently, alarm and despondency were spread in South America by a radio announcer who became bored with the lack of news, and told his listeners that the Moon was about to fall upon the Earth (I understand that the broadcasting company subsequently dispensed with his services). Then, too, there are flying saucers. I do not propose to enter into a discussion about space crockery, but I cannot resist mentioning the occasion in 1958 when I interviewed the late Mr George Adamski, co-author of the classic UFO book *Flying Saucers Have Landed*, on the BBC television programme *Panorama*. Mr Adamski told me that he had been round the Moon, and had seen some dog-like creatures running about on the far side. Strangely enough, these interesting animals have not been confirmed by the astronauts who have since been round the Moon to see for themselves.*

Although the idea of intelligent Moon-men died over a century ago, animals and plants were still considered possible. In fact, the last great advocate of relatively advanced life on the Moon was W. H. Pickering, author of the 1904 photographic atlas as well as many papers about all branches of lunar study.

Between 1919 and 1924 Pickering, observing from the clear climate of Jamaica, carried out a detailed study of the noble crater Eratosthenes which lies at the southern end of the Apennine range. He found some strange dark patches which seemed to show regular variations over each lunation, and although he was sure that tracts of vegetation existed on the Moon he thought it more likely that the Eratosthenes patches, which – he said – moved about and did not merely spread and shrink, were due to swarms of insects.

This startling idea was put forward in Pickering's final paper on the subject, published in 1924. He pointed out that a lunar astronomer of a century earlier would have seen similar moving patches on the plains of North America, due to herds of buffalo, and that the Eratosthenes patches were of about this size; on the other hand they moved more slowly – only a few feet per minute – and it was therefore reasonable to suppose that the individual creatures making them up were smaller than buffalo. Although insects were considered the most likely answer, Pickering's paper contains the following remarkable paragraph:

> While this suggestion of a round of lunar life may seem a little fanciful, and the evidence upon which it is founded frail, yet it is based strictly on the analogy of the migration of the fur-bearing seals of the Pribiloff Islands... The distance involved is about twenty miles, and is completed in twelve days. This involves an average speed of about six feet a minute, which, as we have seen, implies small animals.

* Mr Adamski also told me that the inhabitants of Saturn play a kind of table tennis. I was once a county table tennis player myself, and I would dearly like to play against a Saturnian, but so far it has not been possible to arrange a fixture.

Pickering's idea was that the creatures, animal or insect, travelled regularly between their breeding-grounds and the dark 'vegetation' tracts nearby. His reputation ensured that due attention would be paid to his theory, and he clung firmly to his views right up to the end of his life in 1938. Few people agreed with him, but it was not until the coming of the space-ships that the concept of any kind of lunar life was completely ruled out. Even afterwards I can recall a paper by J. J. Gilvarry, who believed that the lunar maria used to be seas of water, so that the rocks would be of sedimentary type and the colours of the maria due to the remains of long-dead marine organisms.

By now, when the Moon has been so well surveyed and men have been there, no rational person can believe that there is life of any kind. I have seen one suggestion that there might be very primitive organisms in the ice contained in the polar craters, but even if I believed in the presence of ice (which I do not), I would take the idea of life with a very large grain of cosmic salt. There are, of course, a few people with rather extreme views, who have tracked down artefacts on the Moon left there by previous visitors, and who have constructed an elaborate Base on the far side, out of our view. The most curious cult of all is that no men have ever been to the Moon, and the whole programme is an elaborate NASA fake. The sad thing is that these curious people actually believe it; the idea seems to have come from a film, *Capricorn One*, about a bogus landing on Mars. Perhaps it is kindest to say 'No comment'.

The Moon will be inhabited before long, but indigenous life is as unreal as Locke's bat-men or Pickering's insects. It has been left to men of our own world to bring life at last to the barren landscapes of the Moon.

18

The Lunar Base

When I wrote the first edition of my book *Guide to the Moon*, more than half a century ago, lunar travel still lay in the indefinite future, and the idea of a proper Lunar Base sounded more like science fiction than science fact. Things are very different today. A Lunar Base is a real possibility, and could be in place within a decade or two. Much depends upon politics and finance, and the ability of the various nations to work together; I have to admit that looking around at the present world leaders inspires me with a feeling of no confidence at all. But a Lunar Base will be built eventually, and some people now reading this book will live to see it.

Naturally there are financial constraints, and now and then we still hear the old parrot-cry: 'Why spend money on space research, when there is so much to be done at home?' – just as in bygone days there were certainly some who, on principle, objected to the development of the wheel. In fact the space bills are not vast when taken in context with national budgets. I understand that the latest Mars probe cost about one-fortieth as much as a nuclear submarine.

So what will be the advantages of a Lunar Base, and what will be its drawbacks? Let us look briefly at the 'pros' and 'cons'.

Obviously the main drawback is that we can never turn the Moon into a second Earth. If we could wave a wand and provide the Moon with a terrestrial-type atmosphere, the weak gravity would allow it to escape, so that the Moon dwellers – let us call them colonists – will have to stay inside their vacuum suits, inside their space-ships or inside the Base. Moreover, the lunar atmosphere is far too thin to give protection against cosmic rays and harmful emissions from the Sun. True, there is no need to build the Base underground, as was once feared, and the danger from tumbling meteorites seems to be very slight indeed, but at times of high solar activity there will have to be safe, screened areas into which the colonists can retreat.

Plants can never be made to grow in the open, and so far as we can tell there are no minerals valuable enough to be worth mining, though on this score a certain amount of doubt must remain. There is no water supply (unless I am wrong in being sceptical about polar ice), and since the Moon is a quarter of a million miles away there will have to be a transport system which will make the shuttles of 2000 seem very antiquated.

These are a few of the 'cons'. Now let us turn to the 'pros'.

First, the Moon is an ideal site for a physical and chemical laboratory and an astronomical observatory. The low gravity means that all materials are lighter and more manageable than they are on Earth, so that building large, stable structures will be much easier, and so far as stability is concerned the Base will be infinitely superior to any space-station. Large lenses and mirrors will not distort under their own weight to such an extent. Optical telescopes will be used to their full resolving power, because 'seeing' will be perfect all the time, and there is nothing to make the optics deteriorate apart from a certain amount of easily-removed meteoritic dust. Moonquakes will be far too feeble to cause any problems, and of course there is no wind and no 'weather': the Moon's slow rotation will be helpful to astronomical observers.

There is plenty of room on the Moon, and scientific institutions will be able to spread at will, without having to cope with enraged environmentalists or awkward city councils. Radio telescopes can be made to a size which is quite impracticable on Earth, and they can be linked with other radio telescopes, on the Moon itself, on the Earth and in space, to provide very powerful interferometers.

Many sensitive instruments have to be kept at very low temperatures and this normally means cooling by the use of liquid helium. On the Moon this will be unnecessary; with proper shielding, the temperature of the equipment can be kept to any level required.

Astronomically, one major advantage will be the absence of any light pollution or radio interference. The Earth will be too far away to cause any trouble optically, and at radio wavelengths the situation is still good; the far side of the Moon, always turned away from Earth, is completely radio-quiet, and I have no doubt that radio astronomers will make tracks there as soon as they can. The situation on Earth is getting worse all the time, and I remember Sir Bernard Lovell once saying to me that unless something could be done, Earth-based radio astronomy would be a science restricted only to the second half of the twentieth century. To radio astronomers, the South Pole-Aitken Basin will be a paradise. Note, too, that there will be no blocking of radiations with wavelengths either to the long- or the short-wave end of the visible band because of the lack of atmosphere. There will be no point in taking telescopes to the top of Hadley Delta! Solar power will be unlimited, and equipment on the Moon can be controlled from Earth if need be. Undoubtedly there will be many robotic telescopes scattered around the lunar surface.

The Moon itself has no weather in the accepted sense of the term, but of course it will be possible to keep continuous watch on the Earth's weather systems. This will lead to a much better understanding of how our atmosphere behaves, with consequent improvements in weather forecasting. Dangerous storms developing out at sea will also be detected at an early stage, so that anyone living in threatened areas will have time to move away.

The Lunar Physical Laboratory will be one of the major installations, and the airlessness and the low gravity mean that procedures can be carried out which would be impossible on Earth. But in my view, one of the most important of all sections of the Base will be the medical centre. Here the low gravity will be an immense advantage – and heart specialists, for example, are already very interested in this sort of research. It may be that some people who could never live full lives on Earth would be quite happy on the Moon, though I admit that there is the initial problem of getting them there.

As a military outpost, the Moon will fortunately be fairly useless. After all, we already have the weapons to destroy civilization if we feel so inclined, and there is no point in extending the Star Wars system to the Moon.

Space research will benefit. At the moment, vehicles in NEO (Near Earth Orbit) are in real danger from the numerous pieces of débris which cannot be tracked, and this will not be so for the Moon. Very probably a mission to, say, Mars, will start from the lunar station; here again the low escape velocity is a tremendous advantage. For non-fragile materials being sent to Earth it may even be possible to go back to Jules Verne's space-gun, since there is no danger of being burned up by friction against the lunar atmosphere.

Transport between the different lunar centres will be easy enough; in the future there will no doubt be an elaborate system of railways, which will not be delayed by problems such as the wrong kind of rain, the wrong kind of snow, staff shortages, or leaves on the line!

All this may sound futuristic, and in the year 2001 this is true enough, but nothing I have outlined here is beyond the capability of the next generation, provided that we really want to make progress. So far as lunar tourism is concerned, we must wait for the development of really cheap, safe transport vehicles between Earth and Moon. I am confident that tourism will begin well before 2100, and it will be interesting to see who will be the first to climb Mount Huygens, take pictures of the Straight Wall, and explore the shadowed depths of the polar craters.

It is still too early to speculate about what form the Base will take. The old idea of a graceful dome, kept inflated by the air-pressure inside it, may be very wide of the mark, just as the Mir space-station bears little resemblance to the wheel structures envisaged by Wernher von Braun and other twentieth-century pioneers. As a site, we may choose one of the temperate zones, perhaps not too far from the point where Apollo 11 touched down in the Mare Tranquillitatis. We must wait and see. Eventually there will be flourishing colonies – true lunar cities, if you like – and of course all the Bases will be truly international. Living under one-sixth Earth gravity will be by no means uncomfortable, and in any case it is likely that most of those who take up residence on the Moon will return to Earth at regular intervals.

Many of the ideas put forward in this book may prove to be correct. Others will certainly be wrong. If I could come back in, say, the year 2100 I am sure that I would have many surprises. I cannot hope to see very far into the new century, but I am very ready to believe that some of you now reading these words will yourselves go into space – perhaps even to the Moon. It is an exciting prospect. The Moon offers us a challenge, and it is a challenge which Mankind must not ignore.

APPENDIX I

Observing the Moon

When I started paying serious attention to the Moon, more than half a century ago now, it was fair to say that the Moon was still essentially the province of amateurs, and professionals seldom looked at it systematically. The main emphasis was upon mapping. The libration regions, in particular, were poorly known, because they are so foreshortened. Today the situation is very different; the whole of the Moon has been imaged in great detail by the space-probes, so that the rôle of the amateur must also be different.

After Apollo, it was often claimed that further observation of the Moon from the Earth was a waste of time. This brings us back to the situation which prevailed after the publication of the map by Beer and Mädler in 1838 – everything about the lunar surface was already known! It was not until the 'Linné affair' of 1866 that telescopes were once more swung Moonward. And even now there is research available to the amateur, particularly with regard to TLP. Moreover, there is endless pleasure to be gained from simply looking at the Moon through a telescope. A modest instrument will suffice; I began with a 3-inch refractor, which I still have and which I still use. (It cost me £7 10s. This, I must add, was in 1933.)

Many amateurs now use remarkably powerful telescopes, and also electronic devices such as CCDs. I confine myself to a straightforward 15-inch reflector, and in many ways I regard myself as something of an astronomical throwback. What follows may be regarded as slightly old-fashioned; all I am to do here is to give some hints to those who have never before taken serious notice of the Moon.

The first essential is to learn one's way about. My own method, which I recommend to all beginners, was to take an outline map showing a couple of hundred craters, and then to spend my observing time in making at least two drawings of every named formation. One sketch would not suffice, because a crater (or any other feature) can show apparent changes according to the angle at which the sunlight strikes it, and these changes are very marked indeed, as I have stressed over and over again throughout this book. For instance, the huge walled plain Maginus is extremely prominent when near the terminator and filled with shadow, but at or near full moon, when there are virtually no shadows, Maginus is difficult to locate at all.

Full moon, moreover, is the very worst time to start observing, because the whole scene is dominated by the bright ray systems, and the disk appears as a confused medley of bright and darker patches. This was my first mistake, when I set out to learn my way around the Moon. I waited patiently until full moon, looked at my map, and decided to make a start by drawing Ptolemæus, the great walled plain near the very centre of the disk. Not surprisingly, I could not find it. When I looked again during the next lunation, at half-moon, I identified it at once; indeed, I could hardly miss it.

Therefore, the procedure should be to start with the formations near the terminator. Draw them, and keep your results in a file; never throw them away. Next evening the terminator will be further advanced; draw some new craters, and redraw the old ones. Repeat this procedure often enough, and you will soon become familiar with the lunar topography. It took me over a year to complete my first survey, and of course the drawings which I made were very rough and inaccurate, but to me they were invaluable. The programme is laborious, but it does work, and I have never been able to think of any short cut.

My 3-inch refractor was ideal for the purpose, and this is, in fact, the best possible telescope with which to begin. For 'learning' purposes an even smaller telescope is adequate, and it is far better to begin in a modest way, with a small telescope, than to use a powerful instrument at once.

Apart from occultation timings, which come into a different category, it is true that no practical lunar research can be carried out with a very small telescope. At the same time, it is my firm belief that such a telescope is essential if the observer is to turn into a really useful member of a team.

The most common of all faults is that of drawing too large an area on too small a scale. Twenty miles to the inch is a convenient guide, and it is better to be over-generous than parsimonious. I remember that on one occasion I was sent a sketch of the complete Mare Imbrium, made with a 5-inch refractor, in which the Mare itself was about 4 inches across and Plato perhaps a centimetre. Even if the sketch had been accurate (which it was not) it would still have been useless. Far better to select one crater or crater-group only, and make a drawing which is as faithful as possible.

Drawings can be made in various ways. Some observers manage to make representations which are artistic and accurate at the same time, whereas others with less skill – such as myself – prefer to keep to line sketches. From a scientific point of view, the main thing is to make the drawings accurate and easy to interpret.

When the drawing has been completed, add the following data: date, time (using the 24-hour clock, and never using Summer Time), telescope, magnification, name of observer, and seeing conditions. The scale usually favoured for seeing is that introduced by the Greek astronomer E. M. Antoniadi, ranging from 1 (perfect) down to 5 (very poor). If any of these notes are omitted, the drawing will promptly lose most of its value.

One term often encountered is 'colongitude'. This is equal numerically to the longitude of the morning terminator, measured westwards from the centre of the disk. Thus if the colongitude is 270 degrees, the morning terminator is at 90 degrees east – that is to say, the Moon is new. The colongitude is approximately 270, 000, 090 and 180 at new moon, first quarter, full moon and last quarter respectively. (At the morning terminator, the Sun is rising over that part of the Moon; at the evening terminator, it is setting.)

Nowadays, it is probably true to say that lunar drawings are made chiefly for the benefit of the observer himself. But TLP come into a different category, and here the visual observer is in his element, provided that he has an adequate telescope. First, make sure that you know the areas under study really well; the most promising sites are formations such as Alphonsus, Gassendi, Langrenus, Grimaldi and above all Aristarchus, though it would be a great mistake, and would distort the analyses, to concentrate upon these to the exclusion of all others. Really major events are visible in ordinary light, and it now seems that colourless TLP are commoner than red ones.

Use has also been made of the moonblink device, as developed in Britain by P. K. Sartory and V. A. Firsoff. This consists of a pair of filters, one red and one blue, which can be used in rapid succession. As the red filter will tend to suppress a red event and a blue filter will not, the well-known persistence-of-vision effect comes into play, and a faint red patch on the Moon may show up as a 'blinking spot' – hence the name. The device has its limitations, but it is remarkably sensitive, as experiments have shown. The drawing of my own moonblink (fig. 37) will show the main construction. The eyepiece fits in at B; the filters are fitted in at C (red) and D (blue), with a clear area; the filters are rotated by turning the knob (E) and the device screws into the telescope in

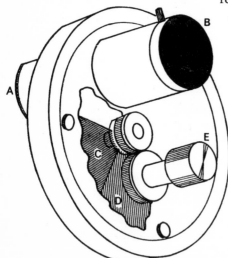

Fig. 37. A moonblink apparatus

the ordinary way (A). Make sure that your filters are of the right type, and are of good optical quality; and of course the blink device cannot be used on a small telescope, as it involves too much loss of light.

TLP work is laborious, and many hours of fruitless searching will pass by before anything positive is seen. (Not so long ago I had a blank period of over two years; I hate to think of the period I spent searching during that time.) Moreover, the utmost care is needed. Many alleged events turn out to be spurious, caused by instrumental defects or by conditions in the atmosphere. Unfortunately, a wrong report will be worse than useless, because it will upset the analyses. If a suspected TLP is seen, check all the features in the surrounding area, and also all the way along the terminator. If other craters show the same effects, you may be sure that your TLP originates in the Earth's air rather than on the Moon.

The Lunar Section of the British Astronomical Association has organized a full TLP network which is proving very successful. Full details of the programme are given in the Section Handbook, listed in the bibliography.

Lunar photography is a fascinating pastime, and amateurs can take superb pictures; indeed, it is true that the best amateur photographs of today are better than the best professional photographs of a few decades ago. Obviously a firmly-mounted, clock-driven telescope is needed for really good results. I am much more of a visual observer than a photographer, so I refer you to the books given in the bibliography.

Finally, there are occultations of stars by the Moon. Accurate timings are still useful, and are not hard to make. All that is really needed is an adequate telescope, plus a good stop-watch; of course the observer must also know his own position on the Earth very accurately. Usually the occultation is virtually instantaneous, though, as I have already said, there are cases of fading immersions – usually (or always?) produced when the occulted star is a close binary. If a star bypasses the upper or lower limb of the Moon, it may be momentarily hidden by mountain peaks along the limb, and these grazing occultations are of special importance, though nearly always the observer has to take a portable telescope to the extremely narrow critical line. One advantage of occultation work is that for the brighter stars, at least, it is within the range of the owner of a comparatively small telescope. Here, too, the Lunar Section of the British Astronomical Association has a wide and effective network of observers.

APPENDIX II
Statistical Data

DISTANCE FROM EARTH	
Mean	238,840 miles or 0.0025695 astronomical units
Maximum	252,700 miles
Minimum	221,460 miles
SIDEREAL PERIOD	27.321661 days
SYNODIC PERIOD	29d 12h 44m 02.9s
AXIAL INCLINATION OF EQUATOR, REFERRED TO THE ECLIPTIC	1°32'
ORBITAL ECCENTRICITY	0.0549
ORBITAL INCLINATION	5°09'
MEAN ORBITAL VELOCITY	2,287 mph = 0.63 miles per second = 3,350 feet per second
APPARENT DIAMETER	max. 33'31", mean 31'5", min. 29'22"
MAGNITUDE OF FULL MOON, AT MEAN DISTANCE	-12.7
MEAN ALBEDO	0.07
DIAMETER	2,160 miles (3,476 kilometres)
MASS	$\frac{1}{81.3}$ Earth = 0.0123 Earth = 3.7 x 10^{-8} Sun
VOLUME	0.0203 Earth
ESCAPE VELOCITY	1.5 miles per second (2.38 km/sec)
DENSITY	3.34 water = 0.60 Earth
SURFACE GRAVITY	0.01653 Earth

APPENDIX III

Eclipses of the Moon 2000–2008

* indicates a total eclipse; otherwise the phase is given.

P indicates that the eclipse is penumbral only.

			DURATION OF ECLIPSE			
		TIME OF MID ECLIPSE	TOTALITY		PARTIAL	
DATE	TYPE	(GMT)	H	M	H	M
2000 21 Jan	*	0445	1	16	3	22
2000 16 July	*	1357	1	0	3	16
2001 9 Jan	*	2022	0	30	1	38
2001 5 July	49	1457			1	19
2001 30 Dec	P(89)	1030				
2002 6 May	P(69)	1205				
2002 24 June	P(21)	2129				
2002 20 Nov	P(86)	0147				
2003 16 May	*	0341	0	26	1	37
2003 9 Nov	*	0120	0	11	1	45
2004 4 May	*	2032	0	38	1	41
2004 28 Oct	*	0305	0	40	1	49
2005 24 Apr	P(87)	0957				
2005 17 Oct	6	1204			0	28
2006 14 Mar	P(100)	2349				
2006 7 Sept	18	1852			0	45
2007 3 Mar	*	2322	0	37	1	50
2008 21 Feb	*	0327	0	24	1	42
2008 16 Aug	81	2111			1	34

APPENDIX IV

Successful Lunar Missions

(a) Russian

NAME	LAUNCH	LUNAS/ZONDS
Luna 1	2 Jan 1959	Passed Moon at 3,700 miles on 4 Jan. Data returned.
Luna 2	12 Sept 1959	Crash-landed in Mare Imbrium, 13 Sept.
Luna 3	4 Oct 1959	Imaged far side of the Moon. Approached Moon to 3,850 miles.
Zond 3	18 July 1965	Approached Moon to 5,800 miles. Images returned.
Luna 9	31 Jan 1966	Landed in Oceanus Procellarum; images returned. Landed 3 Feb; contact maintained till 7 Feb.
Luna 10	31 Mar 1966	Lunar satellite; data returned. Approached Moon to 220 miles.
Luna 11	24 Aug 1966	Lunar satellite; approached Moon to 99 miles. Data and images returned.
Luna 12	22 Oct 1966	Lunar satellite. Images and data returned.
Luna 13	21 Dec 1966	Landed in Oceanus Procellarum, 23 Dec. Images and data returned.
Luna 14	7 Apr 1968	Lunar satellite; approached Moon to 99 miles. Images and data returned.
Zond 5	14 Sept 1968	Went round Moon, approaching to 1,200 miles, and returned to Earth, 21 Sept. Plants, seeds and insects carried.
Zond 6	10 Nov 1968	Went round Moon, approaching to 1,500 miles, and imaged far side. Returned to Earth 17 Nov.
Zond 7	7 Aug 1969	Went round Moon, approaching Moon to 1,250 miles. Images and data returned.
Luna 16	12 Sept 1970	Landed in Mare Fœcunditatis, 15 Sept, collected samples, and returned to Earth 24 Sept.
Zond 8	20 Oct 1970	Circumlunar flight; images of Earth and Moon. Returned to Earth 27 Oct.
Luna 17	10 Nov 1970	Landed 17 Nov. Carried Lunokhod 1 to Mare Imbrium.

Luna 19	28 Sept 1971	Lunar satellite. Contact kept for 400 orbits.
Luna 20	14 Feb 1972	Landed near Apollonius, 17 Feb; collected samples, returned to Earth 25 Feb.
Luna 21	8 Jan 1973	Landed 16 Jan. Carried Lunokhod 2 to area of Le Monnier.
Luna 22	29 May 1974	Lunar satellite. Images and data returned. Contact maintained until 6 Nov 1975.
Luna 24	9 Aug 1976	Landed in Mare Crisium, 18 Aug. Collected samples, returned to Earth 22 Aug.

	Lᴜɴᴏᴋʜᴏᴅs
Lunokhod 1	Carried by Luna 17. Operated in Mare Imbrium for 11 months after arrival on 17 Nov.
Lunokhod 2	Carried by Luna 21. Operated until mid-May 1973, in Le Monnier area.

(b) American

Nᴀᴍᴇ	Lᴀᴜɴᴄʜ	Rᴀɴɢᴇʀs
Ranger 7	28 July 1964	Landed in Mare Nubium, 31 July. Images returned.
Ranger 8	17 Feb 1965	Landed in Mare Tranquillitatis, 17 Feb. Images returned.
Ranger 9	21 Mar 1965	Landed in Alphonsus, 24 Mar. Images returned.

		Sᴜʀᴠᴇʏᴏʀs
Surveyor 1	30 May 1966	Landed near Flamsteed, 2 June. Images returned.
Surveyor 3	17 Apr 1967	Landed in Oceanus Procellarum, 19 Apr. Images and data returned.
Surveyor 5	8 Sept 1967	Landed in Mare Tranquillitatis, 10 Sept. Images and data returned.
Surveyor 6	7 Nov 1967	Landed in Sinus Medii, 9 Nov. Images and data returned.
Surveyor 7	7 Jan 1968	Landed on rim of Tycho, 9 Jan. Images and data returned.

APOLLOS						
NO.	CM NAME	LM NAME	LAUNCH	LAND	SPLASH-DOWN	
7	-	-	11 Oct 1968	-	22 Oct 1968	
8	-	-	21 Dec 1968	-	27 Dec 1968	
9	Gumdrop	Spider	3 Mar 1969	-	13 Mar 1969	
10	Charlie Brown	Snoopy	18 May 1969	-	26 May 1969	
11	Columbia	Eagle	16 July 1969	19 July 1969	24 July 1969	
12	Yankee Clipper	Intrepid	14 Nov 1969	19 Nov 1969	24 Nov 1969	
13	Odyssey	Aquarius	11 Apr 1970	-	17 Apr 1970	
14	Kitty Hawk	Antares	3I Jan 1971	5 Feb 1971	9 Feb 1971	
15	Endeavor	Falcon	26 July 1971	30 July 1971	7 Aug 1971	
16	Casper	Orion	16 Apr 1972	21 Apr 1972	27 Apr 1972	
17	America	Challenger	7 Dec 1972	11 Dec 1972	19 Dec 1972	

APOLLOS					
LAT.	LONG.	AREA	CREW	EVA	SCHEDULE
-	-	-	W. Schirra D. Eisele R. Cunningham	-	Test orbiter (10 days 20h)
-	-	-	F. Borman J. Lovell W. Anders	-	Flight round Moon (6 days 3h)
-	-	-	J. McDivett D. Scott W. Schweickart	-	LM test in Earth orbit (10 days 2h)
-	-	-	T. Stafford J. Young E. Cernan	-	LM test in lunar orbit (8 days 0h)
0°40'N	23°49'E	Mare Tranquillitatis	N. Armstrong E. Aldrin J. Collins	2.2h	Landed; ALSEP
3°12'S	23°24'W	Oceanus Procellarum, near Surveyor 3	C. Conrad A. Bean R. Gordon	7.6h	Landing: ALSEP
-			J. Lovell F. Halse J. Swigert	-	Aborted landing
3°40'S	17°28'W	Fra Mauro formation, Mare Nubium	A. Shepard E. Mitchell S. Roosa	9.2h	Exploration; lunar cart
26°06'N	3°39'E	Hadley Apennine region, near Hadley Rill	D. Scott J. Irwin A Worden	18.3h	Exploration; LRV
8°36'S	15°31'E	Descartes formation, 50 km W of Kant	J. Young C. Duke T. Mattingly	20.1h	Various experiments; LRV
20°12'N	30°45'E	Taurus-Littrow in Mare Serenitatis, 750 km E of Apollo 15	E. Cernan H. Schmitt R. Evans	22h	Geology; LRV

ORBITERS	
	All were successful lunar satellites, returning images, and were deliberately crashed on to the Moon at the end of the mission.
Orbiter 1	Launched 10 Aug 1966. 207 images returned.
Orbiter 2	Launched 7 Nov 1966. 422 images returned.
Orbiter 3	Launched 1 Aug 1967. 307 images returned.
Orbiter 4	Launched 4 May 1967. 326 images returned.
Orbiter 5	Launched 1 Aug 1967. 312 images returned.

(c) Japanese

Hagomoro, launched 24 Jan 1990. Satellite (Hiten) crash-landed on Moon on 10 Apr 1993, near Furnerius.

Hagomoro returned data.

(d) Post-Apollo probes

Clementine, launched 24 Jan 1994. Orbited Moon until mid-May, returning images and data. Now in solar orbit.

Prospector. Launched 6 Jan 1998. Orbiter; search for lunar ice. Deliberately crash-landed in polar crater, 31 July 1999.

(e) Fly-by probes

Various probes have flown past the Moon, returning images: Galileo (8 Dec 1990) *en route* for Jupiter; Cassini (17 Aug 1999) *en route* for Saturn; Shoemaker (as Jan 1998) *en route* for asteroid Eros.

APPENDIX V
Literature and Maps

MAPS

Hatfield, Henry (ed. T. Cook). *The Hatfield Lunar Atlas*. Springer Verlag, 2000. Essential for the amateur. It consists of a series of photographs, covering the whole Moon under several different conditions of illumination, with very clear and accurate maps.

Elger, T. *Lunar Map*. Originally published in 1896 and still available (George Philip & Co.), but unfortunately now printed with north at the top, making it inconvenient for the northern hemisphere.

Moore, P. *Lunar Map*. Essentially a compilation of the maps in this book. Available from the South Downs Planetarium, Chichester, Sussex.

BOOKS

British Astronomical Association, The. *Guide to Observers of the Moon*, 1998.
Montgomery, S. L. *The Moon and the Western Imagination*, University of Arizona Press, 2000
Spudis, P. *The Once and Future Moon*. Smithsonian Institution Press, 1996.
Whitaker, E. *Mapping and Naming the Moon*. Cambridge, 1999.
Williams, D. *To a Rocky Moon*. University of Arizona Press, 1993.

APPENDIX VI

Description of the Surface, and Map

Since this book is intended for observers, it seems only right to include a sectional map, but there are several points which I must make in connection with it. First and foremost, it does not set out to be a precision chart. It is no more than an outline, and is intended to do no more than act as a guide for those people who are trying to find their way about – and want to know which formation is which. I have included all the important names on the

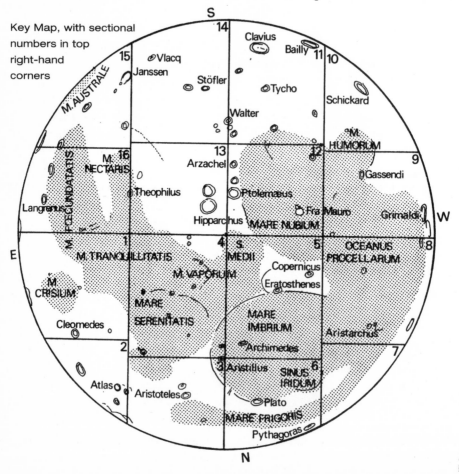

Key Map, with sectional numbers in top right-hand corners

Earth-turned hemisphere, and the IAU directives have been followed throughout except that I have kept to some of the obvious English names for mountain ranges and rills. When the observer has identified all the formations given here it will be high time for him to change to a larger-scale and more accurate map.

There are sixteen sections; 1 to 4 include the First Quadrant, 5 to 8 the Second, 9 to 12 the Third, and 13 to 16 the Fourth. There are bound to be some awkward subdivisions; for instance Archimedes, in the Great Mare Imbrium group, lies in Section 5, while its companions Aristillus and Autolycus are in Section 4 – it is a pity that the boundary had to come just there! However, I have linked each separate map with the sections which adjoin it. The descriptions are very short, and have had to be compressed almost to the point of dehydration, but again I hope that they will serve as a general guide. The depth and diameter measurements are approximate only, but I cannot think that it really matters whether the diameter of – say – Ptolemæus is 92 miles or 93; naturally I have given the figures as accurately as I can.

The map is drawn to mean libration, so that formations such as the Mare Orientale do not appear at all; but I have at least included notes about the few interesting features which are quite invisible except when the libration is at its best for that particular limb.

Finally: though the maps are drawn with south at the top, I have followed the IAU convention with regard to east and west – so that anyone who happens to have an earlier edition of this book will find everything reversed. To recapitulate: Mare Crisium to the *east*, Grimaldi and Riccioli to the *west*.

First (North-East) Quadrant

This quadrant is dominated by Mare areas. It includes the whole of the Mare Crisium and the Mare Serenitatis, as well as the Mare Humboldtianum, Mare Marginis, Lacus Somniorum and Palus Somnii; part of the Mare Frigoris, and practically all of the Mare Tranquillitatis, Mare Vaporum and Mare Smythii. The eastern extensions of the Mare Imbrium – the Palus Nebularum and Palus Putredinis – extend to Sections 3 and 4. Mountain ranges include the Hæmus and Caucasus, as well as part of the Alps and the northernmost extension of the Apennines.

Of craters and walled plains, special note should be made of the dark-floored Julius Cæsar and Boscovich, the brilliant Proclus, Manilius and Menelaus, and the large enclosure of Cleomedes. Aristoteles and Eudoxus form a noble pair; here too are Autolycus and Aristillus, though the third member of the trio, Archimedes, lies in the Second Quadrant (Section 5). The Mare Vaporum is rich in rills; as well as Hyginus and Ariadæus there is the complex

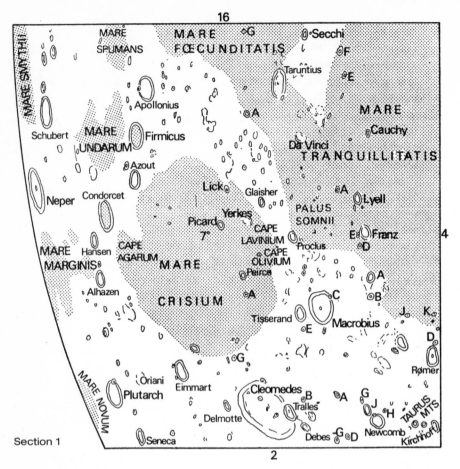

Section 1

system associated with Triesnecker. Last, but certainly not least, the famous (or notorious!) Linné lies in Section 4 on the grey plain of the Mare Serenitatis.

Various space-craft have come down in this quadrant, notably Apollo 17 near the clump of hills known as the Taurus Mountains, and Luna 20 in the highlands of Apollonius, south of the Mare Crisium. Russia's second 'crawler', Lunokhod 2, also carried out its surveys here, near the incomplete crater Le Monnier on the edge of the Mare Serenitatis.

Section 1

ALHAZEN. A crater 20 miles in diameter, near the border of the Mare Crisium. Roughly south of it lies a similar though slightly larger formation, **Hansen**.

APOLLONIUS. A crater 30 miles in diameter, with walls rising to 5,000 feet above the floor. It lies in the uplands south of the Mare Crisium, and south again is the dark patchy area which has been called the '**Mare Spumans**' or Foaming Sea on some maps, though it is certainly not a true Mare.

CAUCHY. An 8-mile crater on the Mare Tranquillitatis; it is bright, and is conspicuous at full moon. There are two fairly long rills nearby, one to either side of Cauchy.

CLEOMEDES. A magnificent enclosure 78 miles in diameter north of the Mare Crisium. The walls average at least 9,000 feet, and there are peaks of even greater altitude. Cleomedes is interrupted by a very deep 28-mile crater, **Tralles**. To the south-east, on the edge of the Mare Crisium, is **Eimmart**, 26 miles in diameter, and other less important formations in this area include **Delmotte** and **Debes**.

CRISIUM, MARE. One of the most conspicuous of the seas, since it is entirely separate from the main Mare system. It measures 280 miles by 350 and is actually elongated in an east-west direction. On it are three craters of some size: **Picard**, **Peirce** and **Peirce B** (the latter was called 'Graham' on the older maps, now renamed 'Swift'). There are also many minor details, including some delicate craterlets closely west of the jutting **Cape Agarum**; these craterlets were first described in detail during the 1930s by the English amateur selenographer R. Barker, and some of them are connected by ridges. Closely outside the Mare in this area is **Condorcet**, a fine regular crater 45 miles in diameter.

FIRMICUS. A 35-mile crater south of the Mare Crisium. Its dark floor makes it conspicuous under any angle of illumination; the walls attain almost 5,000 feet above the interior. Closely outside the north-west wall is a small patch of Mare material. Between Firmicus and the Mare Crisium is **Azout**, 19 miles in diameter and with a low central mountain.

MACROBIUS. A fine walled plain, 42 miles in diameter and with walls reaching 13,000 feet in places. There is a compound central mountain mass of moderate height. Between Macrobius and the Mare Crisium is a smaller crater, **Tisserand**.

MARGINIS, MARE. A limb-sea; it is never well seen from Earth, but Orbiter and Apollo pictures have shown it to be a true Mare of the Crisium type, though less well defined.

NEPER. A deep crater, 70 miles in diameter, between the Mare Marginis and the Mare Smythii. It is, of course, very foreshortened as seen from Earth.

NEWCOMB. A 32-mile crater of considerable depth, west of Cleomedes. Its south wall is interrupted by a craterlet. Newcomb is the northernmost and largest member of a string of three formations. East of it is a smaller but well-formed crater, **Kirchhoff**.

PEIRCE. The second largest crater on the Mare Crisium; it is 12 miles in diameter, and the walls attain about 7,000 feet. North of it is the smaller but probably equally deep **Swift (Peirce B)**, once lettered **Peirce A**.

PICARD. The largest crater on the Mare Crisium; it is 21 miles in diameter, with walls rising to 8,000 feet. There is a central hill. To the west, on the edge of the Mare, lie the two capes **Lavinium** and **Olivium**, together with imperfect craters such as **Lick** and **Yerkes**.

PLUTARCH. A very foreshortened 40-mile crater with a central mountain. On the limb, to the north-east is a small dark area which has been called the 'Mare Novum' or New Sea. Near Plutarch lies the rather smaller **Seneca**, while between Plutarch and Eimmart is the very ill-defined enclosure **Oriani**.

PROCLUS. A brilliant crater west of the Mare Crisium, 18 miles in diameter and 8,000 feet deep. Proclus is one of the brightest points on the Moon, and is the centre of a ray-system; there is a low central mountain. The rays cross the Mare Crisium, but not the Palus Somnii, which is bounded by rays to either side. There are both bright and dusky bands on the inner walls of Proclus.

RØMER. A fine crater, 35 miles in diameter, with high terraced walls rising to over 11,000 feet. The floor contains a particularly large and massive central mountain, on the top of which is a summit pit. From Rømer the so-called Taurus Mountains extend north-east, toward Newcomb and the well-marked Geminus (Section 2), but must be regarded as a hilly upland rather than a true range.

SMYTHII, MARE. Yet another limb-sea of the Crisium type, only well seen from space-probes. A good guide to it is the 46-mile crater **Schubert**, which is not difficult to identify.

SOMNII, PALUS. Really an extension of the Mare Tranquillitatis, bounded on the north-west and south-east by rays from Proclus. The colour is peculiar, and has been described as brownish, greenish or yellowish; it is in any case different from that of the Mare Tranquillitatis itself. On the boundary between the Palus Somnii and the Mare Tranquillitatis are some low-walled craters, notably **Franz** and **Lyell**.

TARUNTIUS. An interesting crater, 38 miles in diameter, with low narrow walls nowhere attaining more than 3,500 feet. There is a central mountain crowned by a pit, and there is a complete inner ring on the floor, so that Taruntius is an excellent example of a 'concentric crater'. Well to the south-west lies **Secchi**, an imperfect but quite conspicuous formation.

TRANQUILLITATIS, MARE. This is one of the major seas, and spreads from Section 1 into Section 4; it joins on to the Mare Serenitatis, Mare Vaporum, Mare Nectaris and Mare Fœcunditatis. The floor is decidedly lighter and patchier than that of the Mare Serenitatis, and the form is less regular. Important craters on it include Cauchy (Section 1) and Maskelyne and Arago (Section 4). It was, of course, in this Mare that Armstrong and Aldrin made their never-to-be-forgotten landing from the lunar module of Apollo 11.

UNDARUM, MARE. The so-called 'Sea of Waves' is a dark patchy area, but not a well-defined sea; it lies near the Mare Crisium, east of Firmicus and Apollonius. It is very dark near full moon, and can be found without difficulty.

Section 2

ATLAS. A magnificent enclosure 55 miles in diameter. The much-terraced walls rise to as much as 11,000 feet above an interior which contains considerable detail – several old rings, some delicate rills, craterlets, and dark patches which seem to vary over the course of a lunation. These changes are, of course, due only to the altering angle of illumination; all the same, they are worth studying. Some way south-east of Atlas is the fine crater **Franklin**, 34 miles in diameter, with walls reaching 8,000 feet; and in the Franklin area are various lesser enclosures such as **Cepheus**, **Œrsted**, **Chevalier** and **Shuckburgh**. Well south of Atlas is the bright, deep crater **Maury**, 11 miles in diameter. The most important neighbour of Atlas is Hercules, with which it forms a noble pair; Hercules is described in Section 3.

BERNOUILLI. A 25-mile crater east of the larger Geminus. The walls of Bernouilli are highest on the east, where they reach up to about 13,000 feet above the floor. To the east is another crater-pair made up of **Berosus** and **Hahn**. Berosus, the larger and more conspicuous, is 47 miles in diameter, with terraced walls.

BERZELIUS. A crater 24 miles in diameter. The floor is rather dark, and includes a central peak; the crater closely east also has a central peak.

CARRINGTON. An unremarkable crater between Messala and Mercurius.

CEPHEUS. The companion to Franklin; it is 28 miles in diameter, and its north-east wall is broken by a bright crater, Cepheus A. Between Cepheus and Franklin there are traces of an old ring.

ENDYMION. An important and interesting crater 78 miles in diameter. Its walls are high, containing a few peaks rising to 15,000 feet, and the floor is very dark, so that Endymion is easy to find under all conditions of lighting. Here, as in Atlas, there are patches which show optical variations over a lunation. The companion to Endymion, the much older and less prominent De La Rue, is shown in Section 3.

GAUSS. Since Gauss has a diameter in the region of 100 miles, it is one of the Moon's grand craters – particularly since the walls are high and continuous. The floor includes a central peak and a long ridge. It is so close to the limb that it can never be properly studied except from space-probe photographs.

GEMINUS. The larger companion of Bernouilli. It is 55 miles in diameter with broad, richly terraced walls attaining 16,000 feet in places. In the centre of the floor there is a rounded hill with a pit on its summit. To the

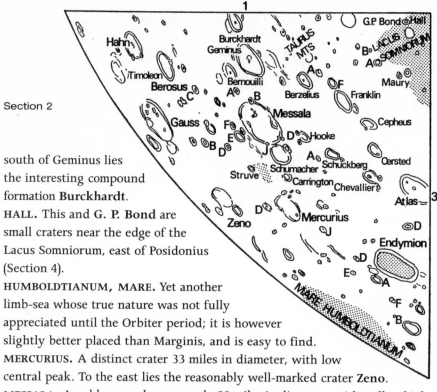

Section 2

south of Geminus lies
the interesting compound
formation **Burckhardt**.

HALL. This and G. P. Bond are
small craters near the edge of the
Lacus Somniorum, east of Posidonius
(Section 4).

HUMBOLDTIANUM, MARE. Yet another
limb-sea whose true nature was not fully
appreciated until the Orbiter period; it is however
slightly better placed than Marginis, and is easy to find.

MERCURIUS. A distinct crater 33 miles in diameter, with low
central peak. To the east lies the reasonably well-marked crater **Zeno**.

MESSALA. An oblong enclosure nearly 80 miles in diameter, with walls which
are broken and generally low. Near it is **Hooke**, only 27 miles across, but
deeper and more distinct. Between Messala and Zeno are **Schumacher**,
smooth-floored and 25 miles across, and **Struve**, a small ring with a central
peak, easy to recognize because it lies on a dark patch.

TAURUS MOUNTAINS. This upland area has been described in Section 1.
It extends from the Berzelius area in the general direction of Rømer.

TIMOLEON. A large ring, 80 miles in diameter, southward along the limb from
Gauss.

Section 3

ALPS. A bright mountain range forming part of the rampart of the Mare
Imbrium, from Plato towards Cassini. Part of it is shown here, and part in
Section 6. The peaks are moderately high; **Mont Blanc** (Section 6), near the
great Valley, rises to 11,800 feet. Note too the small but rather bright crater
Trouvelot, 6 miles in diameter, right in the uplands. The **Alpine Valley** is 80
miles long, and has been described in the text; it is by far the finest formation
of its type on the entire Moon. The rill running down it is a very delicate
object, though of course Orbiter and Apollo photographs show it well.

ARCHYTAS. A bright crater 21 miles in diameter, on the north border of the Mare Frigoris. It has a triple-peaked central mountain, and walls which rise to 5,000 feet above the sunken interior. To the south-east lies a similar but rather smaller crater, **Protagoras**.

ARISTOTELES. A great plain 60 miles across, with walls rising to 11,000 feet. The floor contains many low hills, and particularly notable are the rows of hillocks which radiate outward from the crater itself. Closely outside the east wall is a deep crater, **Mitchell**. Aristoteles forms a notable pair with its slightly smaller neighbour Eudoxus, which lies to the south.

W. C. BOND. A vast enclosure almost 100 miles across, north of Archytas. It is clearly very old, and is in a dilapidated condition, so that it is easy to recognize only under oblique lighting. To the north is **Barrow**, 54 miles in diameter and of the same general type. Many of the walled plains in this neighbourhood have been badly ruined; Meton is another example.

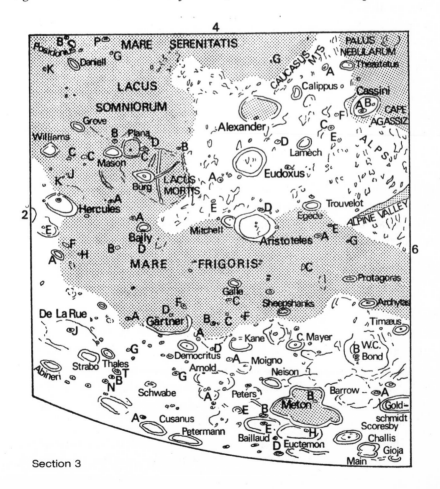

Section 3

BÜRG. A crater 28 miles in diameter, near the edge of the Lacus Mortis. The floor is concave, and the walls rise to 6,000 feet. Bürg is notable both because of its very large central mountain, which includes a summit crater, and because it stands on the eastern edge of a small dark plain which is riddled with rills. Well up to the north is the 12-mile crater **Baily**.

CALIPPUS. A deformed crater 19 miles in diameter, at the northern end of the Caucasus Mountains which separate the Mare Serenitatis from the Palus Nebularum. Closely east of it is **Alexander**, which is 65 miles in diameter, and which has a darkish floor and very low, broken walls.

CASSINI. A peculiar object on the edge of the Palus Nebularum. It is 36 miles in diameter, with very irregular walls; on the floor is a deep, distinct crater, Cassini A, as well as other details. The whole formation was strangely omitted from earlier maps, but there is not the slightest chance that it is of recent origin!

CAUCASUS MOUNTAINS. An important range, forming part of the border between the Mare Serenitatis and the Mare Imbrium (of which the Palus Nebularum is a part). Some of the peaks rise to 12,000 feet. The range extends from this Section into Section 4.

CHALLIS. Challis is 35 miles in diameter. It and its companion, **Main**, form a good example of overlapping; Main, 30 miles in diameter, is the intruding formation. The best 'pointer' in this area is the deep, distinct Scoresby. Challis and Main lie not far from the Moon's north pole, so that they are highly foreshortened and are never well placed for observation from Earth.

DE LA RUE. A large enclosure not far from Endymion, and of about the same size. The walls are, however, very low and broken, and the formation is not distinct. Nearby are two much more prominent craters: Thales, which is a ray-centre, and Strabo.

DEMOCRITUS. A very deep ring 23 miles in diameter, of somewhat distorted shape. It lies in the highland north of the Mare Frigoris, close to the bay of Gärtner. There is a central mountain. To the west are two low-walled formations, **Kane** and **Moigno**. Moigno has a darkish floor. To the north-west of Democritus is the rather low-walled and obscure **Arnold**, 50 miles across.

EGEDE. A peculiar object shaped rather like a diamond; its mean diameter is 23 miles. The floor is dark and the walls are very low. Egede lies roughly between Aristoteles and the Alpine Valley.

EUDOXUS. In many ways Eudoxus is similar to its companion Aristoteles, but it is smaller (diameter 40 miles) and lacks the remarkable radiating rows of hillocks. The walls attain 11,000 feet. Close to Eudoxus is the incomplete formation **Lamèch**.

FRIGORIS, MARE. This is an irregular Mare, and possibly in the nature of an overflow. Its floor is comparatively light and patchy. Craters on it include Galle and Protagoras. The Mare separates the Alpine region from the highlands around the north pole.

GALLE. A small but distinct crater on the Mare Frigoris, north of Aristoteles.

GÄRTNER. A splendid example of a bay. It lies on the border of the Mare Frigoris, and the Mare-material has reduced its 'seaward' wall to such an extent that it is now barely traceable, though the 'landward' rampart is still quite high. Gärtner is 63 miles in diameter, and the floor contains some delicate rills.

GIOJA. A 26-mile crater close to the north pole, and thus very badly placed for observation. It abuts into a larger formation with low walls.

GOLDSCHMIDT. An old ring, 68 miles in diameter, between Barrow and the very prominent Anaxagoras in the Second Quadrant (Section 6). Its walls are low and broken.

HERCULES. The smaller companion of Atlas. Hercules is 45 miles across, and has walls rising to 11,000 feet; these walls are richly terraced, and often appear brilliant. The floor contains one prominent crater as well as a large amount of fine detail. Several TLP have been reported here over the years. Well south of Hercules is **Williams**, just on the Lacus Somniorum.

MASON. This crater is 15 miles in diameter; it forms a pair with Plana (24 miles), not far from Bürg. Both craters have low, broken walls. Plana has the darker floor. South-east of Mason, on the border of the Lacus Somniorum, is a deeper and more distinct crater, **Grove**.

METON. Another of the large enclosures in the north polar region. It lies not far from the distinct Scoresby. Meton is over 100 miles in diameter, but is really a compound formation, made up of several ringed plains which have run together. Between it and the limb, close to its wall, is a smaller but more perfect enclosure, **Euctemon**, and to the north-east is the broken crater **Baillaud**. Near the limb in this area are various other craters of some size, notably **Petermann** and **Cusanus**, all of which are naturally very foreshortened. Smaller rings include **Peters**, **C. Mayer** and **Neison**.

MORTIS LACUS. A small darkish plain near Bürg.

NEBULARUM, PALUS. Part of the Mare Imbrium; the region of Aristillus and Autolycus (Section 6). The name is an old one, but has been omitted on some modern official maps.

PLANA. The companion of Mason, and described with that crater.

SCHWABE. A small, deep crater north of Gärtner, in the highlands.

SCORESBY. A very deep, distinct crater 36 miles in diameter, in the north polar uplands near Challis. It has a twin-peaked central hill. Scoresby is usually easy to recognize, since it is much better-formed than any of its immediate neighbours.

SHEEPSHANKS. A crater just on the Mare Frigoris, well north-west of Aristoteles.

SOMNIORUM, LACUS. This is really a bay leading out of the Mare Serenitatis, but its floor is much lighter and patchier. On its southern border is the great enclosure Posidonius (Section 4), but the 18-mile **Daniell**, the well-formed companion of Posidonius, is in this Section. There is some fine detail on the Lacus, including some delicate rills.

STRABO. A 32-mile crater close to De La Rue. The wall contains some high peaks, and the floor is comparatively smooth. Strabo is the centre of a short and inconspicuous ray system, much less prominent than that of its neighbour Thales.

THALES. Thales, close to Strabo, is 24 miles across, and is a major ray-centre, so that it is very prominent near full moon.

THEÆTETUS. A crater on the Palus Nebularum, near Cassini; it is 16 miles in diameter, and has a low central mountain. It was near Theætetus that Charbonneaux recorded his 'cloud' in 1902, described in the text.

TIMÆUS. A bright crater, 21 miles in diameter, on the north border of the Mare Frigoris, not far from Archytas. The floor contains a double central hill. Timæus is the centre of a minor ray-system, and acts as a good guide to the large, broken enclosure, W. C. Bond.

Section 4

AGRIPPA. A fine crater 30 miles in diameter not far from Hyginus, near the border of the Mare Vaporum. Its walls, which rise to 800 feet, are terraced; and there is a central mountain. It forms a notable pair with its slightly smaller southern neighbour, Godin.

APENNINES. The Apennines are certainly the most spectacular mountains on the Moon, though not the highest; they extend from this Section into Section 5, ending near Eratosthenes. They form part of the rampart of the Mare Imbrium, and make a magnificent spectacle when suitably lit. The loftiest peak in this Section is **Mount Bradley**, near Conon, which rises to 16,000 feet; **Mount Hadley**, at the northern end of the range, is only a thousand feet lower than this, and it was in the foothills of the range in this area that the astronauts of Apollo 15 made their landing. The range ends, to the north, at the jutting **Cape Fresnel**, after which there is a gap between the Mare Imbrium and the Mare Serenitatis until the border is resumed with the Caucasus Mountains. There are various craters in the Apennine uplands, notably the 13-mile **Conon**.

ARAGO. A distant crater, 18 miles in diameter and obviously distorted from the circular form. It lies on the Mare Tranquillitatis; the floor includes a

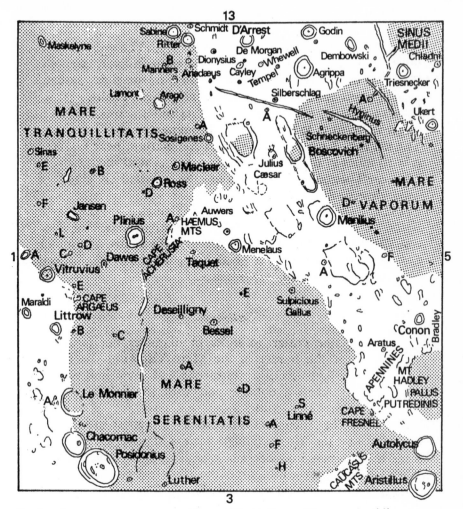

Section 4

central elevation. A low, imperfect ring, **Lamont**, lies to the south-east, and the bright 10-mile crater **Manners** to the south-west. Arago is notable because there are various domes to the west of it. These are among the best examples of domes on the whole of the Moon, and are worth studying; several contain summit pits.

ARATUS. A small bright crater in the Apennines, south of Cape Fresnel.

ARGÆUS, CAPE. A high promontory, guarding the strait between Mare Tranquillitatis and Mare Serenitatis.

ARIDÆUS. A bright 9-mile crater, with a smaller one in contact with it on the north-east. It lies on the border of the Mare Tranquillitatis, and is notable because of the great rill nearby, discovered by Schröter in 1792 and easy to see even with a small telescope. It is over 150 miles long, and has various branches, one of which connects the system with that of Hyginus. The main rill runs across the uplands into the Mare Vaporum.

ARISTILLUS. A splendid crater on the Palus Nebularum, forming a trio with Archimedes (Section 5) and Autolycus. Aristillus is 35 miles in diameter, with walls rising in places to 11,000 feet. The interior details show optical variations over a lunation. The floor includes a fine triple-peaked central mountain.

AUTOLYCUS. The southern companion of Aristillus. It is smaller (24 miles in diameter) but just as distinct, and the walls rise to 9,000 feet. Under high light Autolycus is seen to be the centre of an inconspicuous ray-system, and under a low sun radiating ridges can be made out extending from it.

BESSEL. The largest crater on the Mare Serenitatis; it is 12 miles in diameter, with walls rising to 3,600 feet above the depressed interior. The great ray crossing the Mare Serenitatis passes nearby. To the south-east lies a smaller crater, **Deseilligny**.

BOSCOVICH. A low-walled formation on the border of the Mare Vaporum. It is about 27 miles across, but is irregular in form. It is notable because of the darkness of its floor, which makes it very easy to recognize; the same is true of its neighbour Julius Cæsar.

CAYLEY. One of the bright craters in the uplands between the Mare Vaporum and the Mare Tranquillitatis. It is 9 miles in diameter, and very distinct. The nearby **Tempel** and **De Morgan** are also bright, but only about 5 miles in diameter; **Whewell** is of the same type.

CHACORNAC. A pentagonal ring-plain about 30 miles in diameter, close to Posidonius on the edge of the Mare Serenitatis. There is considerable detail on the floor, including one distinct crater, A.

DAWES. A 14-mile crater between the Mare Serenitatis and the Mare Tranquillitatis; it is somewhat deformed. Two dusky bands run from the small central peak up the inner west wall.

DIONYSIUS. A brilliant crater, 12 miles in diameter, on the edge of the Mare Tranquillitatis not very far from Ariadæus. It stands on a light area, and is very conspicuous under high illumination. It is yet another crater with dark bands running up its walls.

GODIN. A crater 27 miles in diameter, slightly deformed, but with a central hill. It is the southern member of the Godin-Agrippa pair. Between it and the Sabine-Ritter pair lies **D'Arrest**. The low-walled **Dembowski** lies some way west of Godin and Agrippa.

HÆMUS MOUNTAINS. These mountains form part of the border of the Mare Serenitatis, and separate it from the Mare Vaporum. They are not lofty, but some of the peaks rise to about 8,000 feet. The glittering Menelaus lies on the edge; so does the much smaller and less bright **Auwers**.

HYGINUS. A depression about 4 miles in diameter, notable because of its association with the famous crater-rill which has been fully described in the text. North of it lies Hyginus N, and here too is an interesting spiral mountain, **Mount Schneckenberg**, which requires a fairly high power to be well seen. Some branches of the Hyginus rill-system join up with the system of Ariadæus.

JULIUS CÆSAR. An imperfect, very dark-floored enclosure between the Mare Tranquillitatis and the Mare Vaporum, not far from Boscovich. This is one of the darkest patches on the entire Moon. Outside its east wall is a crater-valley.

LE MONNIER. A fine example of a bay; it lies on the border of the Mare Serenitatis, and is 34 miles in diameter. Of the seaward wall, only a few mounds now remain; the floor holds little visible detail. It was in this region that Russia's Lunokhod 2 came down – and where it still remains!

LINNÉ. This celebrated formation, on the Mare Serenitatis, has been fully described in the text.

LITTROW. A 22-mile crater between Le Monnier and Cape Argæus, on the edge of the Mare Serenitatis. The walls are of some height, but are broken in places. Apollo 17 came down in the Taurus-Littrow area.

MANILIUS. A splendid crater 25 miles in diameter, on the border of the Mare Vaporum. Its walls are brilliant, so that Manilius is very prominent near full moon. The walls are terraced; there is considerable interior detail, including a central mountain.

MARALDI. A low-walled crater near Rømer (Section 1).

MASKELYNE. A 19-mile crater on the Mare Tranquillitatis, not very far from the Apollo 11 landing site. The walls have inner terraces, and there is a low central peak. To the west is a conspicuous little crater, Maskelyne B.

MENELAUS. A brilliant crater in the Hæmus Mountains, striking under a high light even though it is only 20 miles in diameter. The walls rise to 8,000 feet above a floor which contains a peak not quite centrally placed.

PLINIUS. A superb crater 'standing sentinel' on the strait between the Mare Serenitatis and the Mare Tranquillitatis. It is 30 miles across, but appreciably distorted from the circular form. The central structure takes the form of a twin crater. Plinius has high terraced walls, and is conspicuous under all conditions of illumination.

POSIDONIUS. A walled plain 62 miles in diameter, adjoining Chacornac and lying on the borders of the Mare Serenitatis and the Lacus Somniorum. The ramparts are rather low and narrow, and the floor is crowded with detail, including a nearly central craterlet, Posidonius A. On the Mare, to the west, is the craterlet **Luther**.

PUTREDINIS, PALUS. Part of the Mare Imbrium, north-west of Mount Hadley. It spreads on to Section 5.

RITTER. This and its neighbour Sabine form a striking pair on the Mare Tranquillitatis, again in the general area of Apollo 11. Ritter is very slightly the larger, and is about 19 miles across. Nearby is a small but bright crater, **Schmidt**, and closely north of Ritter are two more bright craterlets, making almost perfect twins of the sort so common on the Moon. Both Sabine and Ritter have central peaks.

ROSS. A crater with a central peak. It is 18 miles in diameter, on the Mare Tranquillitatis. To the south-west of Ross lies **Maclear**, rather dark-floored and 11 miles across, west of which is a fine long rill running to, and beyond, Sosigenes A.

SABINE. The companion of Ritter, and described with it.

SCHNECKENBERG, MOUNT. The strange spiral mountain, described with Hyginus.

SERENITATIS, MARE. One of the most perfect of the regular seas. It covers an area of 125,000 square miles, slightly more than that of Great Britain. It is bordered, in part, by the Caucasus Mountains and by the Hæmus Mountains, ending at **Cape Acherusia**. The floor is relatively smooth, but contains several craterlets of which the most prominent are Bessel, Deseilligny and Luther; of course Linné also lies on the Mare. There are many prominent ridges, some of which seem to be the walls of ghost-craters, and the controversial bright ray runs right across the Mare, from Menelaus through to Bessel, Luther and on into the highlands beyond.

SILBERSCHLAG. A small, bright crater 8 miles in diameter, near Ariadæus. It is not unlike Cayley, though slightly smaller.

SINAS. This and its smaller companion, E, lie on the Mare Tranquillitatis, well north of Maskelyne.

SOSIGENES. A 14-mile crater to the east of Julius Cæsar; it has a small central hill, and its walls are bright. To the south-east is Sosigenes A, which lies on a long rill running from Maclear, and which is connected to Sosigenes itself by a ridge.

SULPICIUS GALLUS. An extremely bright crater, 8 miles in diameter, and with walls which rise to about 8,000 feet. It lies just on the Mare Serenitatis, in the foothills of the Hæmus Mountains. There is a small central peak.

TAQUET. (Sometimes spelled 'Tacquet'.) Another bright craterlet just on the Mare Serenitatis, in the Hæmus region; it is 6 miles across.

TRANQUILLITATIS, MARE. A major sea, extending on to Sections 1 and 13. It is less regular in outline than the neighbouring Mare Serenitatis, and has a lighter, patchier floor. On it are various craters; some have already been described, and others include the regular **Sinas** and the low-walled **Jansen**.

TRIESNECKER. A crater in the area of the Mare Vaporum and of the Sinus Medii, which is described in Section 12; west of it is the distinct crater **Chladni**. Triesnecker is 14 miles across, and is notable because of the very complex rill-system to the east of it. The chief rills are visible with a small telescope under good conditions.

UKERT. Another 14-mile crater at the edge of the Mare Vaporum, north of Triesnecker. Here too there are rills, though the system is a minor one and does not rival that of Triesnecker. Ukert itself has rather bright walls.

VAPORUM, MARE. One of the minor seas, but notable because of its darkness and because of the many interesting objects nearby – such as Boscovich, Manilius and the rill-systems of Ariadæus, Hyginus, Triesnecker and Ukert. A small part of it extends on to Section 5.

VITRUVIUS. An interesting formation, just on the Mare Tranquillitatis near Mount Argæus. The walls are bright, but the floor is decidedly dark, with a low central peak. Well to the south-west is Jansen, which is 16 miles across and has very low walls, rising to only about 3,000 feet above its darkish floor – which is scarcely depressed below the general level of the Mare. About the same distance south-west of Jansen is the small, well-marked crater Sinas. North-east of Vitruvius, in the uplands, lies **Maraldi**, which is reasonably well formed.

Second (North-West) Quadrant

This is the 'sea' quadrant, and there is little true upland. Most of the area is covered by the two greatest maria on the Moon – the Mare Imbrium and the Oceanus Procellarum, together with various minor seas such as the Sinus Æstuum, the Sinus Roris and the lovely Sinus Iridum or Bay of Rainbows. Only along the limb do we find tremendous walled plains, of which Pythagoras and Xenophanes are good examples. On the Mare surface lie three of the most famous of all lunar craters: Copernicus, Archimedes and Aristarchus, while the dark-floored Plato is to be found on the border of the Mare Imbrium.

The southern part of the Apennine range lies in this quadrant, ending near the majestic Eratosthenes; there are also the less lofty Carpathian Mountains, near Copernicus, and the Jura Mountains, bordering Sinus Iridum. There are also some lofty peaks near the limb.

Section 5

ÆSTUUM, SINUS. A conspicuous dark plain east of Copernicus, bordered by the southernmost extension of the Apennines. Eratosthenes lies at its boundary. The floor is relatively devoid of important craters.

APENNINES. This superb range stretches up from Section 4, ending at Eratosthenes. This part of it contains the highest peak, **Mount Huygens** (18,000 feet); there is also the triangular mountain mass **Mount Wolf,** at least 12,000 feet, and **Mount Ampère**, around 11,000 feet. There are a few low-walled, rather distorted craters in the Apennine uplands, such as **Marco Polo**, 12 miles across with a darkish floor, and **Serao**.

ARCHIMEDES. The largest of the craters on the Mare Imbrium; it forms a trio with Aristillus and Autolycus (Section 4). Archimedes is 50 miles in diameter, with walls around 4,300 feet above the slightly sunken floor; the rampart includes a few peaks reaching to at least 7,000 feet, but in general the wall has been much reduced. The floor is dark and (by lunar standards) very smooth, with no vestige of a central mountain.

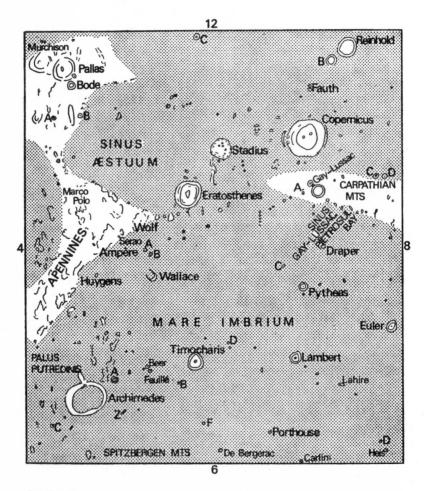

Section 5

BEER. An 8-mile crater on the Mare Imbrium, between Archimedes and Timocharis. It has an almost identical twin, **Feuillé**, closely north-west of it. Like those of Messier and Messier A, the relative sizes of Beer and Feuillé seem to vary; these apparent changes are due only to optical effects, but they are interesting nonetheless. Between Archimedes and the Beer-Feuillé pair is a small deep crater, A (now named Bancroft), with a central peak.

BODE. An 11-mile bright crater in the highlands separating the Sinus Medii from the Sinus Æstuum. Its walls reach to 5,000 feet above the floor. Bode lies close to the semi-ruined crater-ring Pallas, and is conspicuous near full moon; it is the centre of a minor ray-system.

CARLINI. A small, rather bright crater on the Mare Imbrium. It is 5 miles in diameter, with a small central hill; its isolated position makes it conspicuous.

CARPATHIAN MOUNTAINS. A range forming part of the border of the Mare Imbrium, in the Copernicus area. It cannot compare with the Apennines, and is rather discontinuous, though it extends for a total of between 200 and 250 miles. Its highest peaks attain 7,000 feet, but most of the mountains are much lower than this. The western end of the range stretches into Section 8.

COPERNICUS. The 'Monarch of the Moon'. It has been described in the text, and no more need be said here except to repeat that it and its rays dominate this whole part of the lunar surface.

ERATOSTHENES. This also ranks as one of the most perfect craters on the Moon. It is 38 miles in diameter, and extremely deep, with central elevations and much floor-detail. It marks the southern end of the Apennine range.

EULER. A minor ray-centre on the Mare Imbrium, and therefore easy to find under a high light. Euler is a well-marked crater 19 miles in diameter, with a central peak and walls which show some inner terracing.

FAUTH. A double craterlet south of Copernicus; its form makes it easy to identify. This whole area is dominated by ridges and crater-chains radiating from Copernicus. Fauth was splendidly shown on the celebrated 'Picture of the Century' of Copernicus and its environs, taken by Orbiter 2 – the best lunar probe picture taken up to that time, though surpassed many times since.

GAY-LUSSAC. An irregular 15-mile crater in the Carpathian Mountains, north of Copernicus. Its floor contains some fine detail, including delicate rills. To the west, in the direction of Tobias Mayer, are two bays in the Carpathian range which have been named **Sinus Gay-Lussac** and **Pietrosul Bay**. Immediately south-east of Gay-Lussac itself is the smaller, deeper crater Gay-Lussac A, and to the north of the bays are two or more small craterlets, **Draper** and Draper C.

HEIS. A small craterlet of the Mare Imbrium, between De L'Isle (Section 8) and Caroline Herschel (Section 7). Heis is very minor, but is easy to find because of its isolated position.

IMBRIUM, MARE. The greatest of all the regular seas. Most of it lies in this Section, but it also extends into Sections 6 and 8. It has been fully described in the text, but it is worth repeating that in area it is larger than Great Britain and France combined.

LA HIRE. A bright 5,000-foot mountain on the Mare Imbrium, north-west of Lambert. It has a summit craterlet.

LAMBERT. A crater only 18 miles across, and not bright; but it is easy to find, owing to its position on the Mare Imbrium. In place of a central mountain, it has a central crater – a type of feature not uncommon on the Moon.

MURCHISON. The companion of Pallas, on the edge of the Sinus Medii. It has been badly distorted, and is clearly very old, so that its walls are now low and broken. It is about 35 miles in diameter.

PALLAS. Pallas, adjoining Murchison, is rather smaller (diameter 30 miles) but more complete, even though its walls have been broken by mountain passes. The central peak still exists. On the other side of Pallas is the smaller but much deeper crater Bode.

PUTREDINIS, PALUS. Part of the Mare Imbrium, to the west of Archimedes.

PYTHEAS. A very bright crater on the Mare Imbrium, with terraced walls and a central hill. It is a minor ray-centre, and is so conspicuous that it is surprising to find that its diameter is a mere 12 miles. Well to the south lie Draper and Draper C. Rays from Copernicus cross this whole area.

REINHOLD. A 30-mile crater on the Oceanus Procellarum, south-west of Copernicus; its walls rise to 9,000 feet in places. Closely north-east of it is Reinhold B, which has low walls and is about 15 miles across, with a darkish floor.

STADIUS. The celebrated 'ghost', on the edge of the Sinus Æstuum. It forms a triangle with Copernicus and Eratosthenes. It has been described in the text.

TIMOCHARIS. An interesting crater on the Mare Imbrium, roughly west of Archimedes. It is 25 miles in diameter, with broad terraced walls reaching to 7,000 feet. Like Lambert, it has a central crater. Timocharis is the centre of a rather faint ray-system, and is easy to identify.

WALLACE. An incomplete crater between Archimedes and Eratosthenes.

Section 6

ANAXAGORAS. A crater 32 miles in diameter, in the north polar uplands and closely west of Goldschmidt (Section 3), which it distorts. Anaxagoras is well formed, with high walls and a central mountain. It is extremely bright, and is the centre of a major ray-system, so that it is easy to find under all conditions of illumination. It is a pity that it is not better placed for observation from Earth.

ANAXIMANDER. A walled plain, 54 miles in diameter, with a good deal of floor-detail (though no central peak) and walls which rise in places to 9,000 feet. Adjoining it to the north-east is the smaller but almost equally deep **Carpenter.** The limb-formations in this area are of great interest, and there are some splendid walled plains which come into view only under conditions of extreme libration, so that they cannot be shown on these maps. The newcomer to lunar work is often surprised to see them – as I originally was!

BIANCHINI. A bright-walled, somewhat irregular crater in the Jura Mountains, bordering Sinus Iridum. It has a central peak, and is usually easy to find. Its diameter is about 25 miles. To the east of it, also in the Jura range, is the very low-walled, irregular enclosure **Maupertuis.**

BIRMINGHAM. A very large enclosure, about 66 miles across, in the highlands north of the Mare Frigoris, and east of Fontenelle. It has been

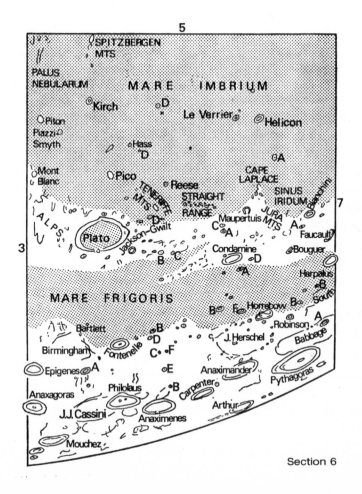

Section 6

much distorted, and is crowded with detail, but it is not too easy to identify unless really well placed with regard to illumination.

CONDAMINE. A 30-mile crater in the foothills of the Jura Mountains, on the border of the Mare Frigoris. Its walls are broken by passes and one distinct crater. To the north-west is a distinct, deep crater, Condamine A.

EPIGENES. A 30-mile crater with broad walls, south of Anaxagoras.

FONTENELLE. A bright walled plain with a central crater. It lies on the northern edge of the Mare Frigoris, and is easy to find on account of its brilliance. Close by it is the famous 'Mädler's Square', described in the text; and north of Fontenelle there is an ill-formed plain, **J. J. Cassini**, bounded by irregular ridges.

FRIGORIS, MARE. The western part of the Mare Frigoris extends into this Section, from Section 3. The general aspect is the same as that of the eastern part.

HARPALUS. A conspicuous crater 32 miles across, near the borders of the Mare Frigoris and the Sinus Roris. It has an asymmetrical floor-mountain. Not far from it are two more distinct, rather deep craters, **Foucault** and **Bouguer**, both much smaller than Harpalus. Incidentally, Harpalus was the crater selected as a rocket launching-site in the famous film *Destination Moon*, produced in the early 1950s when space-travel was still regarded as something of a joke!

HELICON. This and Le Verrier form a conspicuous pair on the Mare Imbrium, near the Sinus Iridum. Helicon is 13 miles in diameter, Le Verrier 11; Helicon has a central craterlet, Le Verrier a central peak. Both have moderately high walls. Oddly enough Helicon is always easy to find, but Le Verrier becomes so obscure near full moon that it is difficult to locate at all.

HERSCHEL, JOHN. A ridge-bordered enclosure on the edge of the uplands bordering the Mare Frigoris to the north. It is about 90 miles across, and its floor is very rough. To the south-west are several distinct craters, including **Horrebow** and **Robinson**.

IMBRIUM, MARE. Part of the vast Mare Imbrium extends into this Section. The general aspect is the same as that of the area in Section 5.

IRIDUM, SINUS. The lovely 'Bay of Rainbows'; one of the most spectacular sights on the Moon near sunrise, when the floor is in shadow and the **Jura Mountains** are illuminated, so that they seem to protrude into the blackness and give the 'jewelled handle' effect. From the Mare, the level of the Bay gradually slopes down to the extent of 2,000 feet. The seaward wall has almost vanished; its site is now marked only by a few very low, irregular ridges and one or two small craterlets. The capes to either side of it are **Heraclides** and **Laplace**.

KIRCH. A bright 7-mile crater on the Mare Imbrium, north of the Spitzbergen Mountains. Well to the north-west lies a somewhat smaller crater-pair.

LE VERRIER. This is the companion of Helicon, and has been described with it.

NEBULARUM, PALUS. Part of the Mare Imbrium, extending into this Section from Section 3.

PHILOLAUS. A very deep, prominent crater 46 miles across, with walls rising to 12,000 feet above a floor which contains a good deal of detail. Unusual colour effects have been reported here. **Anaximenes** adjoins Philolaus to the west; it is slightly the larger of the two, but not nearly so deep or conspicuous. To the east of Philolaus is the ill-formed J. J. Cassini, already described.

PICO. A bright mountain on the Mare Imbrium, south of Plato. It is triple-peaked; the maximum height is about 8,000 feet. The area between Pico and Plato is occupied by a very large ghost-ring; it was once known as Newton, though the name has now been transferred to a crater in the opposite part of the Moon. The 'ghost' has now (at my suggestion) been renamed **Bliss**, after Nathaniel Bliss, the 18th-century Astronomer Royal. The bright **Teneriffe Mountains**, some of which are almost as lofty as Pico, also lie near the boundary of the destroyed ring, and may once have formed part of its wall though more probably they are of considerably later date.

PITON. Another isolated mountain, 7,000 feet high, and with a summit craterlet. It lies well to the south-east of Pico, and is always easy to find, because of its brightness. Between the two peaks, rather closer to Piton, is the bright 6-mile crater **Piazzi Smyth**, and there are various other mountains and small craterlets in the area.

PLATO. Hevelius' 'Greater Black Lake'; the 60-mile crater noted for the darkness of its floor, which makes it easy to recognize under any conditions of lighting. Many TLP have been reported here. To the north-west is the well-marked deep crater Plato A, named on some old maps 'Jackson-Gwilt'.

PYTHAGORAS. One of the grand craters of the Moon, with its lofty, continuous walls and central mountain mass. It is, unfortunately, too foreshortened to be seen properly from Earth, but space-probe photographs bring it out in its true guise.

SOUTH. This is a ridge-bounded enclosure about 60 miles across. It and its larger and even more ruined neighbour **Babbage** lie close to Pythagoras, and are full of detail, but their walls are so broken and discontinuous that they are often difficult to recognize. The nearby **Robinson** is only 17 miles in diameter, but is generally easier to locate.

SPITZBERGEN MOUNTAINS. A series of bright little hills north of Archimedes (Section 5). They lie on the edge of a very obscure ghost-ring which is now traceable only because of a slight difference in the hue of what must once have been its floor.

STRAIGHT RANGE. A remarkable line of peaks, rising to a maximum of 6,000 feet, on the Mare Imbrium between Plato and Cape Laplace. The peaks are less brilliant than Pico or Piton, but are still quite bright. The range is curiously regular, and there is nothing quite like it anywhere else on the Moon.

TENERIFFE MOUNTAINS. These little peaks lie near Pico, on the border of 'Bliss', the ghost-ring between Pico and Plato. They have been described with Pico.

Section 7

CLEOSTRATUS. A large, well-formed enclosure very close to the limb, not far from Xenophanes; there are various other rings in the libration zone. Cleostratus has steep, rather narrow walls.

GERARD. Another large, well-formed limb feature, with a long ridge running down its floor. Further on the disk is a 14-mile crater, **Harding**, with low walls; north-east of Harding is another small crater, **Dechen**. Other limb-craters in this area are **La Voisier** and (not shown on a mean libration map) **Régnault, Galvani** and various others. The crater **Naumann**, some way from La Voisier, has fairly bright walls.

HERSCHEL, CAROLINE. An 8-mile crater on the Mare Imbrium, forming a triangle with Carlini (Section 5) and De L'Isle (Section 8). Its isolated position makes it conspicuous, particularly as its walls are rather bright.

IRIDUM, SINUS. A small part of the Sinus, including Cape Heraclides, appears in this Section, but most of it is in Section 6.

LICHTENBERG. A small crater on the Oceanus Procellarum. It is 12 miles in diameter, and is a minor ray-centre, though the rays are very short. Under high light, Lichtenberg appears as an ill-defined whitish patch. Reddish 'events' have been reported here now and then ever since the time of Beer and Mädler. Between Lichtenberg and the limb is a well-formed crater, **Ulugh Beigh**, which is 30 miles across and has a central mountain.

MAIRAN. A fine, well-formed crater in the Jura uplands. Its walls are lofty, but there seems to be no central mountain. North of it, also in the Jura uplands, is the ill-defined depression **Louville**, which is not hard to identify because of its dusky floor.

ŒNOPIDES. A prominent crater near Cleostratus and Xenophanes. It is 42 miles in diameter, and has high walls, broken in the south-west by a craterlet. There are some minor features on the floor, though, like many of its companions in this part of the Moon, it lacks a central peak. Immediately east of it lies Babbage, which has already been described (Section 6), and to the south-west lies the bright, regular crater Œnopides A.

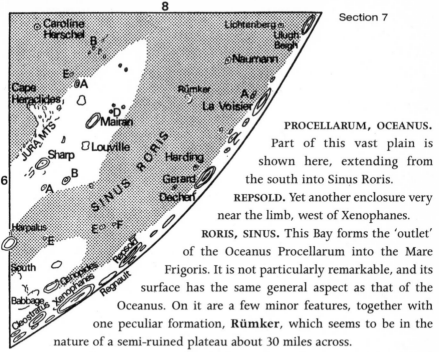

8

Section 7

PROCELLARUM, OCEANUS. Part of this vast plain is shown here, extending from the south into Sinus Roris.

REPSOLD. Yet another enclosure very near the limb, west of Xenophanes.

RORIS, SINUS. This Bay forms the 'outlet' of the Oceanus Procellarum into the Mare Frigoris. It is not particularly remarkable, and its surface has the same general aspect as that of the Oceanus. On it are a few minor features, together with one peculiar formation, **Rümker**, which seems to be in the nature of a semi-ruined plateau about 30 miles across.

SHARP. A deep crater 22 miles in diameter, in the Jura Mountains and surrounded by high peaks. The floor includes a small central peak. Two small, fairly deep craters, A and B, lie to the north-west.

ULUGH BEIGH. This large crater lies near Lichtenberg, and has been described with it.

XENOPHANES. A grand 67-mile crater, with a massive, elongated, central mountain crowned by a craterlet. The walls are lofty and terraced. However, it is very badly placed, and is much too near the limb to be well seen even when libration is at its maximum.

Section 8

ARISTARCHUS. The brightest formation on the Moon — and the most 'active', since gaseous emissions have been proved, and the area including Aristarchus, Herodotus, Prinz and the Harbinger Mountains has been responsible for more than half the number of TLP reported over the years. As early as 1911 R. W. Wood, in the United States, took some ultra-violet photographs which led him to believe that a small area near Aristarchus was covered with a sulphur deposit, or at any rate something quite unlike the surrounding regions; this is still sometimes referred to as 'Wood's Spot'. The extreme brilliance of Aristarchus makes it obvious under any conditions, even when illuminated only by earthshine.

9

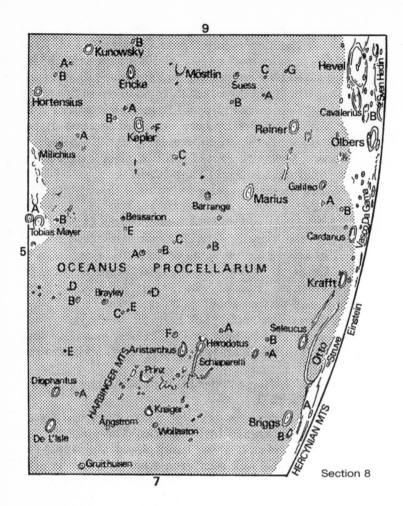

Section 8

7

BESSARION. A bright 6-mile crater on the Oceanus Procellarum. It has a central hill, and a dark band up the inner south-west wall. North of Bessarion is a smaller but equally bright crater, Bessarion E, sometimes called **Virgil**.

BRAYLEY. Another crater on the Oceanus Procellarum, similar to Bessarion but rather larger (diameter 10 miles). It too has a low central hill and dusky radial bands in its interior. It is a member of a curved line of craterlets, of which B, south-east of Brayley itself, has rather bright walls.

BRIGGS. A 33-mile crater on the Oceanus Procellarum, east of Otto Struve. It is well marked, and easy to locate. Ridges connect it with Seleucus.

CARDANUS. This and **Krafft** form another notable pair in the limb-region at the very edge of the Oceanus Procellarum. Cardanus has a diameter of 32 miles and continuous walls attaining 4,000 feet. It has a central mountain, and there are numerous craterlets on its floor. To the south-east, between

Cardanus and Reiner, lies the bright little crater **Galileo** (more properly, Galilaei), which is fairly bright – yet a 9-mile formation is surely inadequate to honour the first man to undertake serious, regular telescopic studies of the Moon! As has been noted in the text, Riccioli, who named this crater in his 1651 map, disliked Galileo as being a protagonist of the heretical theory that the Earth moves round the Sun instead of vice versa, so that when the names were being allotted Galileo was treated very badly indeed.

CAVALERIUS. The northern member of the chain which includes Hevel as well as Lohrmann, Riccioli and Grimaldi (Section 9). Cavalerius is well formed, with a diameter of 40 miles and a central ridge on its floor. The walls rise to 10,000 feet above the interior. It is a fine object when on the terminator.

DE L'ISLE. A 16-mile crater, forming a pair with the less regular 13-mile **Diophantus**, to the south. Both craters have central peaks. Various domes lie in the area, and are worth studying.

ENCKE. A crater which may be regarded as the 'twin' of Kepler; but it is a dissimilar twin, since it is far less bright and is not a ray-centre. The diameter is 20 miles, and the walls are rather low. There is no central peak, but a ridge lies on the floor. To the west may be seen the unremarkable formation **Möstlin**.

GRUITHUISEN. A bright crater on the Oceanus Procellarum, 10 miles in diameter. The area between it and Aristarchus is of great interest. There are the small, bright craters **Ångstrom** and **Wollaston** and the rather larger, less regular **Krieger**, together with rills and domes. The incomplete ring **Prinz** has domes on its floor, and there are others nearby. The **Harbinger Mountains** do not form a proper range, but are made up of groups of hills, the highest of which rises to 8,000 feet; all the same, it is possible that the Harbingers once formed part of the border of the Mare Imbrium, perhaps connecting the modern Carpathians with the Jura Mountains.

HERCYNIAN MOUNTAINS. A limb-range near Otto Struve. Some of the peaks here may exceed 7,000 feet in altitude.

HERODOTUS. The companion-crater to Aristarchus. It is 23 miles across, and has a darkish floor, in striking contrast to the brilliance of its neighbour. Its walls rise to about 4,000 feet, but the shape of the crater is not quite regular. Issuing from it is the grand valley usually called **Schröter's Valley** in honour of its discoverer – though this may lead to some confusion, as the crater named after Schröter is a long way away (Section 12). The Valley has been described in the text.

HEVEL. A magnificent 70-mile crater in the Grimaldi chain. Its walls are almost linear in places, but rise to 6,000 feet above a decidedly convex floor which contains a central mountain. Hevel is noted for the system of rills inside it, superbly shown on space-probe pictures but very prominent even from Earth.

HORTENSIUS. A deep 10-mile crater in the Oceanus Procellarum, well formed and with rather bright walls. To the north may be seen a group of domes; some of the domes have summit pits.

KEPLER. In diameter (22 miles) Kepler is very like Encke, but it is far brighter, and is the centre of one of the most prominent ray-systems on the Moon. There is a central mountain, and the walls are so heavily terraced that they seem to be double in places. To the south-east is a small, deep craterlet, Kepler A. Like Aristarchus, Kepler has interior radial bands, though they are not nearly so pronounced as those in Aristarchus or a few other craters. Kepler has been the site of several reported lunar events.

KUNOWSKY. A small crater, 12 miles across, south-east of Encke. Its floor contains a rather low central ridge.

MARIUS. A well-marked crater 22 miles in diameter, on the Oceanus Procellarum. It has a low central hill, as well as bright streaks on its floor. Well to the south-west is its 'twin', **Reiner**, slightly smaller (diameter 20 miles) with brightish walls but a dark floor.

MAYER, TOBIAS. A crater in the Carpathian Mountains, 22 miles across, and with a central hill; adjoining it to the east is Tobias Mayer A, which is smaller but which also has a central hill. There are various rills and domes in this region.

MILICHIUS. A small bright crater 8 miles in diameter, on the Oceanus Procellarum, west of Copernicus. West of it lies a magnificent dome with a summit pit, and there are various other domes not far off.

OLBERS. A 40-mile crater north-west of Cavalerius; between the two is a small crater, B. Olbers is a major ray-centre, and like all ray-craters is very bright. Various other formations beyond it come into view under conditions of extreme libration.

PRINZ. The famous partial ring in the Harbinger Mountains. It has been described with Aristarchus.

PROCELLARUM, OCEANUS. The vast 'Ocean of Storms' has an area of two million square miles, much larger than our Mediterranean, but it is not one of the well-formed circular seas; its surface is lighter and patchier than that of the Mare Imbrium. It extends on to Sections 5 and 9, and joins the Mare Nubium, Mare Humorum, Mare Imbrium and Sinus Roris. Under adverse libration the Oceanus spreads almost to the limb of the Moon, though none of it lies on the far hemisphere.

REINER. Reiner is not unlike Marius, and has been described with it. Various small craters lie on the Oceanus to the south-west; **Suess** is the largest of them.

SELEUCUS. This may be regarded the twin of Briggs, to which it is connected by ridges. It is 32 miles in diameter, with terraced walls rising to 10,000 feet above an interior which contains a central peak. North-east of Seleucus is the distinct 18-mile crater **Schiaparelli**; a light streak runs south-westward from it, so that Schiaparelli is easy to find under a high light.

STRUVE, OTTO. A vast enclosure made up of two old rings, each about 100 miles across, which have now merged; the west wall is really a ridge parallel with the mountains beyond. In 1952 I discovered a vast crater, only visible under ideal conditions of libration and lighting, beyond Otto Struve and the extreme limb; it has high walls, a central crater and a long ridge on its floor. It had not been previously recorded because it is so seldom visible from Earth in its true guise. It was originally named 'Caramuel', but the IAU Commission altered this name to **Einstein**. Photographs of Einstein from space-probes show it to be a truly magnificent formation.

SVEN HEDIN. A large 60-mile irregular formation, with broken walls, between Hevel and Cavalerius on the one side and the limb on the other. There is considerable floor-detail. Sven Hedin is an interesting structure; Olbers, to its north, is the best guide to it.

VASCO DA GAMA. Though 50 miles in diameter, and with a central ridge rising to a peak at its mid-point, Vasco da Gama is too foreshortened to be well seen from Earth. It lies north of Olbers and west of Cardanus and Krafft; Einstein is situated to its north-west.

Third (South-West) Quadrant

The Moon's Third Quadrant includes most of the tremendous Mare Nubium as well as the Mare Humorum and a small part of the Oceanus Procellarum, but much of the quadrant is composed of upland. There are mountains along the limb, and of these the so-called Cordillera and Rook ranges are really parts of the complicated system of the Mare Orientale – something which could never be known before the age of space-probes. There are also the D'Alembert Mountains, but there is some risk of confusion here, as the name has been left off the official maps and the name transferred to a crater on the Moon's far side. Also omitted is the name of the 'Percy Mountains', formerly used for the lofty uplands bordering the Mare Humorum to the west.

Major craters and walled plains in this quadrant include Grimaldi, Riccioli, Schickard, the Walter and Ptolemæus chains, Bailly, Clavius, and of course Tycho. There are notable rill-systems associated with Hippalus, Hesiodus and Sirsalis, and among other features we have the celebrated plateau Wargentin and the remarkable fault miscalled the Straight Wall. Various lunar vehicles have come down in this quadrant, and the region of Mare Nubium in which Apollo 12 landed in 1969 is often called the Mare Cognitum or Known Sea.

Section 9

AGATHARCHIDES. An irregular formation about 30 miles across, with walls which are of fair height in places (up to 5,000 feet) but which have been almost levelled in others. There is the remnant of a central mountain. The border of the Mare Humorum between Agatharchides and Gassendi has been destroyed – assuming, of course, that it ever existed!

BILLY. This and **Hansteen** make up a pair on the edge of the southern part of the Oceanus Procellarum. Each is about 32 miles in diameter, with walls rising to between 3,000 and 4,000 feet, but Billy is notable because of the darkness of its floor, making it readily identifiable under high illumination.

CORDILLERA MOUNTAINS. As we now know these and the **Rook Mountains** make up part of the ring boundaries of the Mare Orientale, so that they extend well on to the far hemisphere.

CRÜGER. A 30-mile crater with a very dark floor, resembling that of Billy. Near it are two small dark plains which have been called the Mare Veris and the Mare Æstatis, though they do not seem to merit separate names. Crüger itself can always be found easily under high illumination; it is of the same type as Plato, and is almost exactly half the size.

D'ALEMBERT MOUNTAINS. High peaks on the limb, again associated with the Mare Orientale complex.

DAMOISEAU. A very complicated formation east of Grimaldi, on the edge of the Oceanus Procellarum, made up of several old crater-rings. The total diameter is between 20 and 30 miles.

DARWIN. A large, semi-ruined enclosure west of a line joining Byrgius (Section 10) to Crüger; Crüger, with its dark floor, is a good guide to it. The floor of Darwin contains various rills as well as a large and important dome. To the north-east of Darwin is the deep 15-mile crater **De Vico**; the nearby De Vico A lies at the southern end of the great Sirsalis Rill.

EUCLIDES. A remarkable little crater close to the Riphæan Mountains (Section 12). It is only 7 miles in diameter, and 2,000 feet deep, but it is surrounded by an extensive bright nimbus which makes it very prominent. There are several much smaller bright craterlets nearby.

FLAMSTEED. A 60-mile ghost ring with very low walls, but quite distinct under high light.

FONTANA. A 30-mile crater east of Crüger, with low but bright walls and a central hill. There are various rills in this region, which is of course close to the extensive Sirsalis rill-system.

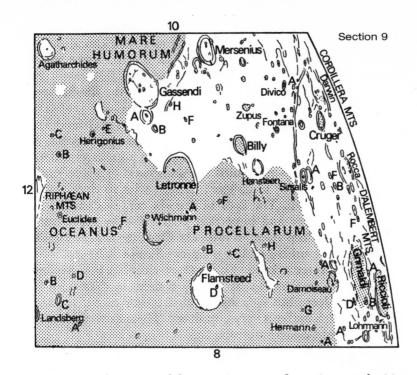

GASSENDI. This is one of the most important formations on the Moon. It is 55 miles in diameter, and lies on the north border of the Mare Humorum. The wall is reasonably high to the east and west, but to the south it contains numerous passes, and has obviously been badly damaged by the Mare material, while to the north the wall has been broken by a prominent, well-formed crater, Gassendi A (named 'Clarkson' on some former maps). The floor of Gassendi includes a central peak, and a magnificent system of rills, shown to advantage on probe pictures but also easily seen from Earth. Gassendi is one of the most 'event-prone' areas on the Moon, and various transient phenomena have been seen there in recent years – which is hardly surprising, since it lies on the edge of a regular sea (Humorum) and is also so rich in rills.

GRIMALDI. The famous dark-floored walled plain near the west limb. It is 120 miles in diameter, so that it is one of the largest enclosures on the Moon, and its floor is so dark-hued that Grimaldi is always unmistakable. The walls are discontinuous, but include some peaks which exceed 8,000 feet. The ramparts are extremely complex, and include hills, ridges, and rills near their foot. The chief feature on the floor is the well-marked crater B. Many TLP have been reported in and near Grimaldi, and gaseous emissions have been detected spectroscopically.

HANSTEEN. This is the companion-crater to Billy, and has been described with it. Unlike Billy, it has a fairly bright floor.

HERIGONIUS. A 10-mile crater on the Mare, north-east of Gassendi. It has rather bright walls, and a central hill.

HUMORUM, MARE. A small part of this interesting Mare is shown in the present Section, but most of it lies in Section 10.

LETRONNE. A good example of a bay. It has a diameter of 70 miles, and borders the Oceanus Procellarum, north-west of Gassendi. Its north wall has been destroyed, and the floor is fairly smooth, though it contains the wreck of a central peak. North of Letronne is a small, distinct craterlet, A.

LOHRMANN. A 28-mile crater lying between Hevel (Section 8) and Grimaldi, so that it is a true member of the celebrated chain. It has a darkish floor, on which is a central hill, and there are many rills nearby. Running obliquely in this region is a curious valley, not well shown on many of the probe pictures, but quite striking under suitable illumination; it was studied by the Japanese astronomer Miyamori, and is known unofficially as the **Miyamori Valley**. East of Lohrmann, on the Oceanus, is the bright 10-mile crater **Hermann**.

MERSENIUS. An important and interesting crater, 45 miles across, near the border of the Mare Humorum. Its walls are terraced, rising to about 8,000 feet in places, and the floor is markedly convex; Hevel is another large formation with this peculiarity. Mersenius is associated with an extensive rill-system; some of the rills lie on the Mare Humorum, others to the south of Mersenius in the direction of Liebig and De Gasparis.

PROCELLARUM, OCEANUS. A small part of the Oceanus extends into this Section, and includes the craters Wichmann, Flamsteed and Hermann as well as the great bay Letronne.

RICCIOLI. The smaller companion of Grimaldi. It is 100 miles in diameter, and has one patch on its floor which is almost as dark as any area in Grimaldi. The interior contains much fine detail.

ROCCA. A large crater, 60 miles in diameter, with irregular walls. It lies south of Grimaldi and north of Crüger.

SIRSALIS. A 20-mile crater which overlaps its slightly larger neighbour Sirsalis A, so that the two form a striking pair similar to that of Steinheil in the Fourth Quadrant (Section 14). Sirsalis, the intruding formation, is much deeper than its twin. Nearby is the famous rill, visible with any small telescope when well placed, which extends from the border of the Oceanus Procellarum southward as far as Byrgius (Section 10).

WICHMANN. A bright 8-mile crater on the Oceanus Procellarum. It is associated with a ghost crater which is even more ruined than that of Flamsteed.

ZUPUS. A very low-walled formation only about 12 miles across, south of Billy. It is easy to find, because its floor, like those of Billy and Crüger, is extremely dark.

Section 10

BYRGIUS. A low-walled and rather obscure enclosure about 40 miles in diameter, not far from the limb. It is easy to find because the small crater on its eastern crest, Byrgius A, is a ray-centre, and so is prominent under high light. Byrgius lies north-west of Vieta.

CAVENDISH. A formation 32 miles in diameter. Its walls are of fair height, attaining 7,000 feet in places, but are disturbed by smaller craters. It lies south-west of Mersenius (Section 9) and west of the Mare Humorum. Between it and the Mare border are various small craters, of which the most prominent are **Liebig** and **De Gasparis**.

DOPPELMAYER. Another splendid lunar bay, this time on the edge of the Mare Humorum. The 'landward' wall is quite high, and there is a central mountain, but the 'seaward' wall has been so ruined that it is now very low and discontinuous, with wide gaps. Adjoining it is **Lee**, another incomplete formation which has been ruined by the Mare lava, but which is much less impressive than Doppelmayer. **Palmieri**, roughly between Doppelmayer and Vieta, is a curious enclosure whose floor is crossed by several rills.

EICHSTÄDT. A 32-mile regular-walled crater north-west of Byrgius, and not at all prominent. The Mare Orientale lies beyond.

FOURIER. A 36-mile crater close to Vieta. Its walls are terraced, and the floor contains a central crater in lieu of a central peak.

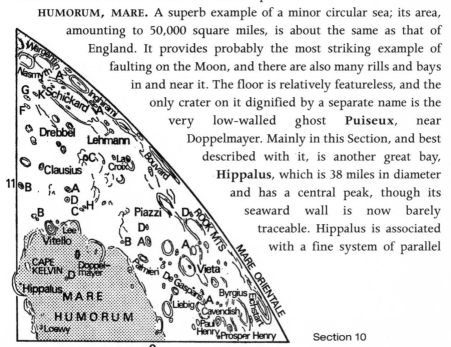

HUMORUM, MARE. A superb example of a minor circular sea; its area, amounting to 50,000 square miles, is about the same as that of England. It provides probably the most striking example of faulting on the Moon, and there are also many rills and bays in and near it. The floor is relatively featureless, and the only crater on it dignified by a separate name is the very low-walled ghost **Puiseux**, near Doppelmayer. Mainly in this Section, and best described with it, is another great bay, **Hippalus**, which is 38 miles in diameter and has a central peak, though its seaward wall is now barely traceable. Hippalus is associated with a fine system of parallel

Section 10

rills. To the south of Hippalus is the jutting **Cape Kelvin**; to the north a much less prominent bay, **Loewy**.

INGHIRAMI. A beautiful walled plain 60 miles in diameter, between Schickard and the limb. It has high terraced walls, with peaks rising to 10,000 feet, and a central mountain. From it a ridge runs toward the even larger **Bouvard**, 80 miles across, and which has a central ridge rising to a peak at its mid-point.

ORIENTALE, MARE. It lies to the west of Eichstädt and Rocca, and since it is quite invisible except under favourable libration it cannot be shown on the main map. Not until space-probe pictures of it were obtained did anyone realize how significant and important it is. It has been described in the main text, and there is no point in giving topographical details here, because it is so hard to see and because it extends right on to the far hemisphere. The **Rook Mountains** and **Cordillera Mountains** form its outer and inner ring-wall respectively, which was something else which we could not possibly know when we found it; there is also a fine, regular, central-peaked crater, Schlüter, virtually inaccessible to Earth-based observers. Mare Orientale has proved to be immensely complex. Further on the disk are the large formations **Lagrange** and **Piazzi**.

SCHICKARD. This is one of the Moon's greatest walled plains, with a diameter of 134 miles. The walls are rather low, averaging less than 5,000 feet, and with its highest peaks rising to just over 8,000 feet. The floor contains some dark areas, as well as various hills and craterlets. Adjoining it to the north-west is the 28-mile **Lehmann**, whose floor is connected by passes with that of Schickard. Not far off is the 18-mile, well-formed crater **Drebbel**, as well as the small but quite prominent **Clausius** and the ill-defined, rather dark-floored **Lacroix**. The most interesting of Schickard's neighbours is, of course, Wargentin, described separately, while Phocylides and Nasmyth (Section 11) are also members of the Schickard group.

VIETA. A 50-mile crater. Its walls are irregular in height, but in places rise to 15,000 feet; there is a minor central peak.

VITELLO. This is a splendid example of a concentric crater – even though the complete inner ring is not quite concentric with the main wall. Vitello is 30 miles cross, and has a central peak crowned by a craterlet. It lies on the border of the Mare Humorum, east of Doppelmayer and Lee, and its seaward wall has been clearly reduced by the Mare lava, though elsewhere the rampart rises to over 4,000 feet above the interior.

WARGENTIN. This is one of the most remarkable formations on the Moon, and represents the only example of a really large, well-preserved lunar plateau. It is 55 miles in diameter, so that in size it is the equal of Copernicus, and adjoins Schickard. There are various hills and ridges on the plateau surface. There is a 'wall' in places, but the whole floor is raised by about 1,400 feet. Despite its unfavourable position close to the limb, it is easy to find, and is well worth careful study.

Section 11

BAILLY. The largest formation on the Earth-turned hemisphere which is officially classed as a walled plain. Its area is more than half that of the Mare Humorum, but it has a light floor. It has been described as 'a field of ruins'; even though peaks in its walls rise to about 14,000 feet, the height of the rampart is very irregular. The floor is crowded with detail, notably one large, well-formed crater, B. West of it, even nearer the limb, is **Hausen**; and along the limb toward the south pole there are various quite major formations such as **Legentil, Drygalski, Cabæus** and **Malapert**, which are difficult to see because of extreme foreshortening. Before the space-probe pictures became available, this whole region was very imperfectly mapped. Beyond Bailly lie some peaks which have been called the **Dörfel Mountains**, though it is now known that they do not make up a lofty, continuous range, as was once thought likely.

BLANCANUS. This and its companion **Scheiner** are rather dominated by their vast neighbour Clavius, but both are major formations in their own right. Blancanus is 57 miles in diameter, Scheiner 70; both have lofty walls, rising to 12,000 and 15,000 feet respectively; both have much interior detail – a nearly central craterlet in the case of Scheiner. Close to the limb near here are several prominent craters, **Wilson, Kircher, Bettinus, Zucchius** and **Segner**, which are more or less lined up and were probably formed along the same line of weakness in the lunar crust. All five are between 40 and 50 miles across, with lofty, continuous walls. The area between this crater-line and Scheiner seems to be occupied by what looks like an old ring, though its walls have been completely destroyed.

CAMPANUS. A well-formed, 30-mile crater at the edge of the Mare Nubium and the Palus Epidemiarum. It is the twin of Mercator, but its floor is lighter, though still on the dusky side. There is a central hill. Various rills run between Campanus and Hippalus, associated with the Hippalus system. To the south-west is **Dunthorne**, which has rather broad walls.

CAPUANUS. This is an extraordinary formation. It is 35 miles across, and lies on the edge of the Palus Epidemiarum; its floor has been to some extent flooded, and appears darkish, while the walls have been disturbed on the seaward side and badly ruined in places. What makes Capuanus so notable is the fact that at least eight major domes lie on its floor; some of them are visible with small telescopes. There is no other known case of a large crater

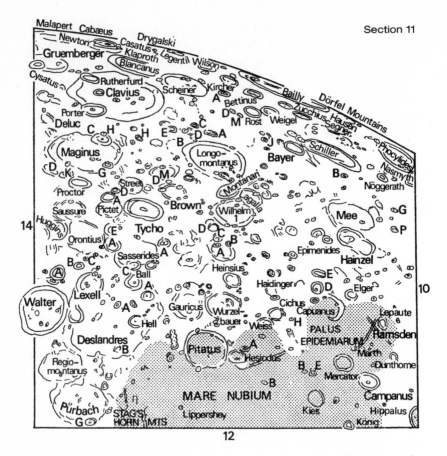

so rich in domes. Closely outside Capuanus is an imperfect formation, **Elger**.

CASATUS. This and **Klaproth** form another example of overlapping craters – this time on a grand scale, since Casatus, the smaller of the two, is a full 65 miles in diameter, with high walls. Klaproth is shallower, with a much smoother floor. Not far from Klaproth, to the west, are several well-formed craters. Drygalski, already referred to, lies in the libration region beyond Casatus.

CICHUS. A prominent 20-mile crater east of Capuanus, just beyond the border of the Mare Nubium. On its western crest is a well-formed crater, Cichus G, about 5 miles in diameter. To the south is the distinct Cichus A, and to the north of Cichus is **Weiss**, which is so broken that it gives the impression of being an enclosure bounded by irregular ridges rather than a true crater. There are various rills in this area, no doubt associated with the Hesiodus system.

CLAVIUS. Apart from Bailly, Clavius is the largest of the so-called walled plains on the Earth-turned hemisphere, since it is 145 miles in diameter, with mighty walls rising to over 12,000 feet. Every lunar observer knows it well. The north-east wall is broken by a large crater, **Porter**, and there is a chain of craters arranged in an arc across the floor, of which **Rutherfurd** is the largest. When on the terminator, Clavius is distinctly visible with the naked eye, and is easy to find telescopically under any conditions of illumination.

DELUC. An unremarkable 28-mile crater south-east of Maginus.

DESLANDRES. A huge enclosure west of Walter and Regiomontanus; on older maps it was called Hörbiger (a name due to the German selenographer Philipp Fauth, a devotee of the extraordinary ice theory proposed by Hans Hörbiger, and described in the text). Deslandres is distorted in outline. The regular crater **Hell** lies inside it; on the border are **Ball**, which is 25 miles across and has high terraced walls, and the 39-mile **Lexell**, whose north wall has been reduced to such an extent that Lexell now resembles a bay opening out of Deslandres, even though it retains the wreck of a central peak. In the north part of the floor of Deslandres is a low-walled crater, B.

EPIDEMIARUM, PALUS. A conspicuous dark plain extending out of the Mare Nubium. Mercator and Campanus lie on its borders. It is notable mainly because there are many rills in and near it, known as the Ramsden system even though the crater of **Ramsden** itself is only 15 miles in diameter. In complexity, the Ramsden system rivals that of Triesnecker. Closely west of Ramsden is another small, unremarkable crater, **Lepaute**.

GAURICUS. An irregular enclosure with a diameter of about 40 miles, and walls which are uneven in height. It is a member of the Pitatus group; Wurzelbauer is the third member.

GRUEMBERGER. This and its smaller companion **Cysatus** lie south-west of Clavius, and belong to the Moretus group (Section 14).

HAINZEL. A curious formation, made up of two rings which have coalesced; the north-south diameter is 60 miles. Under oblique lighting it is conspicuous, but it is hard to find near full moon. It lies north of Schiller. East of it is a small crater, **Epimenides**, and to the south is a large ruined enclosure, **Mee**, with low walls.

HEINSIUS. A very peculiar structure between Tycho and Capuanus, rather closer to Tycho. The north wall is quite high, but in the south the rampart has been broken by three considerable craters, one of which really lies on Heinsius' floor. The diameter of Heinsius itself is about 45 miles.

HELL. A 20-mile crater with a low central hill, lying near the western edge of the great enclosure Deslandres.

HESIODUS. This is the companion of Pitatus, to which it is connected by passes in their common wall. It is 28 miles across, and its walls have been somewhat reduced by the Mare lava; the floor contains an almost central crater. From Hesiodus a famous rill runs across to the mountain arm north of Cichus; it is easy to see with a small telescope when well placed. Other, less prominent rills are associated with it.

KIES. This cannot be termed a ghost-ring, but it lies on the Mare Nubium, and its walls are now very low, nowhere exceeding 2,500 feet; the floor is flooded by lava. West of it lies a superb example of a dome with a summit pit, and there are other domes in the area. Other smaller craters nearby are **König**, Kies A, and Agatharchides A, in which is a dark interior radial band, and which lies on one of the Hippalus rills.

KLAPROTH. This large crater has been described with its companion, Casatus.

LIPPERSHEY. A 4-mile crater west of the Stag's-Horn Mountains, which actually intrude into this Section from Section 12.

LONGOMONTANUS. A very large enclosure, 90 miles in diameter, with complex walls and considerable detail on its floor. It is easy to find, though not nearly so prominent as Clavius. North of it, between Longomontanus and Wilhelm I, is **Montanari**, with two distinct craters on its west wall, which is common to another rather ill-defined walled plain, **Lagalla**.

MAGINUS. With a diameter of 110 miles, Maginus is one of the grandest walled plains on the Moon, and would seem even more striking were it not so close to the even more majestic Clavius. Maginus has walls of irregular height, and a rather rough floor. To the south-west the rampart is broken by a 30-mile crater, Maginus C; another crater of about the same size, **Proctor**, lies outside the north wall. Oddly enough Maginus is difficult to identify near the full moon – even if it does not vanish entirely, as some books maintain.

MERCATOR. The twin of Campanus, on the edge of the Palus Epidemiarum. It has a dark floor, with walls rising in places to 5,000 feet; there is only a trace of a central peak, but the floor contains some detail, including a delicate rill. To the south-west is **Marth**, which has a complete inner ring and is an excellent example of a 'concentric crater'.

NEWTON. This is generally regarded as the deepest of the lunar walled plains. It appears to be compound, but is so foreshortened that to Earth-based observers it is a difficult object to study. It lies south-west of Moretus (Section 14) and east of the Casatus-Klaproth pair. Between Newton and Moretus is another crater, **Short**, which also lies in Section 14.

NUBIUM, MARE. Part of the Mare Nubium is shown here, but most of it lies in Section 12.

ORONTIUS. An irregular formation about 52 miles in diameter, north-east of Tycho. It is one of a group which includes Miller and Nasireddin in Section

14, and **Huggins** and **Saussure** in this Section. Orontius has been disturbed by Huggins; the whole area is crowded with detail, but includes no particularly notable features.

PHOCYLIDES. A most interesting formation, 60 miles in diameter. It is a member of the Schickard group (Section 10), which also includes **Nasmyth** (between Phocylides and Wargentin) and Wargentin itself, as well as Phocylides, to the north-west. There is considerable detail on the floor of Phocylides, and there are indications of a major step-fault there. Between Phocylides and the limb lies a regular crater, **Pingré**; and between Phocylides and Mee there is another crater, **Nöggerath**, which is unremarkable but distinct.

PITATUS. A grand formation 50 miles across. It lies on the border of the Mare Nubium, and gives the impression of a large lagoon; its walls have been badly damaged, and are very low in places. Passes connect the floor with that of Hesiodus. Pitatus has no true central peak, but there is a hill not quite in the middle of the floor. The other members of the Pitatus group are Gauricus and Wurzelbauer.

PURBACH. This great ring-plain, 75 miles in diameter and with walls rising to 8,000 feet in places, is a member of the chain which includes Regiomontanus and Walter. There is considerable detail on the floor. The outline of Purbach is not entirely regular, and the northern part of the rampart has been disturbed by later outbreaks. To the north-west lies a fairly regular crater, **Lacaille**, which comes between this Section and Section 14; it is also adjacent to Blanchinus (Section 14).

REGIOMONTANUS. A distorted formation between Purbach and Walter. Its east-west diameter is 80 miles, but the north-south diameter is only 65 miles; one has the impression that the whole crater has been 'squashed' between Purbach and Walter. There is abundant floor-detail, including some peaks near the centre. The walls are of irregular height, with some peaks reaching 7,000 feet. Regiomontanus touches both Purbach and Deslandres.

SASSERIDES. An irregular enclosure north of Tycho, with a diameter of about 60 miles. Its north wall has been largely destroyed by four smaller craters, and the floor includes numerous pits and hills. To the south-west is Sasserides A, which has a central peak, and between Sasserides and Orontius is a regular formation, Orontius A.

SCHEINER. This great plain may be regarded as the twin of Blancanus, though it is the larger and deeper of the two. It is described with Blancanus.

SCHILLER. A compound formation in the area between Schickard and Clavius. It is 112 miles long, but only 60 miles wide at its broadest point, and is the result of the fusion of old rings. The floor contains some ridges and pits. The region between Schiller, Segner and Phocylides is probably a very old ring whose walls have been destroyed. Closely outside Schiller is the well-formed **Bayer**, 32 miles across, and with high terraced walls attaining 8,000 feet in places.

TYCHO. This crater, 54 miles in diameter, is in a class by itself, as it is at the centre of by far the greatest ray-system on the Moon; it has been fully described in the text, and the last vehicle of the Surveyor series now stands on its outer slopes. Tycho's neighbours include **Pictet**, **Brown** and **Street** as well as Sasserides and the Orontius group.

WALTER. A majestic, complex walled plain 90 miles in diameter, with an asymmetrically-placed interior mountain and several considerable craters on its floor. It is the senior member of the trio which includes Regiomontanus and Purbach.

WILHELM I. A 60-mile walled plain with walls which are irregular in height, but which contain a few peaks reaching 11,000 feet above the floor. Outside it, to the south-west, is the irregular, rather pear-shaped Lagalla; to the north-west is a deepish crater, D, and beyond this, in a northerly direction, is Heinsius. Wilhelm I is easy enough to find, as it lies not far from Tycho, but it does become very obscure under high illumination.

WURZELBAUER. This is one of the Pitatus group, and is about 50 miles in diameter, but its walls are irregular, and very low in places. The floor contains a mass of complex detail.

ZUCCHIUS. This is one of the chain which includes Segner, Bettinus, Wilson and Kircher. It has been described with Blancanus. Between Zucchius and Longomontanus lie two unremarkable but distinct craters, **Weigel** and **Rost**.

Section 12

ALPETRAGIUS. Though only 27 miles in diameter, Alpetragius is a splendid sight when observed under good conditions. It lies closely outside the walls of Alphonsus and Arzachel, and is distinguished for its very high terrraced walls, which exceed 12,000 feet in places, and for its enormous central mountain, which is rounded and crowned by two summit pits.

ALPHONSUS. Little more need be said about this great formation here, since it has been so fully described in the main text. It achieved particular notoriety in 1958 when Kozyrev saw a red event inside it, but for many years before that it had been regarded as a possible site of mild activity. It is the middle member of the Ptolemæus chain, and is slightly distorted in form; on the floor there is a minor central mountain and a mass of detail, including a fine system of rills. It also contains the remains of Ranger 9, though no telescope on earth will show it! Obviously, Alphonsus should be kept under close surveillance, since further TLP may occur in it at any time.

ARZACHEL. The southern member of the Ptolemæus chain. It is 60 miles in diameter, and has high walls, reaching 13,500 feet in places; the floor includes an elongated central mountain and a deep, prominent crater, Arzachel A, well over to the west of the peak. The gradation in type from Ptolemæus to Alphonsus and then Arzachel is very interesting and significant. There can be no doubt that all three were formed by the same process, which adds force to my contention that there is no basic difference between a 'walled plain' such as Ptolemæus and a true crateriform structure such as Arzachel – or, for that matter, Alpetragius.

BIRT. A very interesting little crater 11 miles in diameter, on the Mare Nubium to the west of the Straight Wall. Its south-east wall is disturbed by a smaller crater, A, and at the junction of the walls there is a still smaller formation. The walls and profile of Birt are irregular, and there are two dusky bands running across the floor to the western wall; these show optical changes over a lunation. Closely outside, to the west, is the famous rill, which is in part crateriform. Birt may be found even under high light, when the Straight Wall is completely invisible.

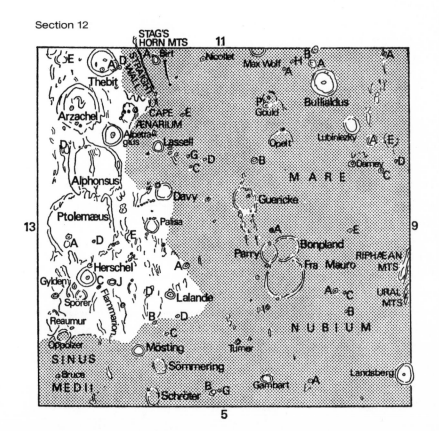

Section 12

BONPLAND. This is a member of the Fra Mauro group on the Mare Imbrium, and is described with Fra Mauro.

BULLIALDUS. A particularly fine crater; it has been described as a miniature Copernicus, though it is not a ray-centre. The diameter is 39 miles. The massive walls rise to an average height of over 8,000 feet above a floor which includes a complex central mountain group; the inner slopes of the rampart are superbly terraced. Bullialdus is, in fact, one of the most perfect of all formations of its type. To the south are two prominent deep craters, Bullialdus A and B, with König (Section 11) some way to the west of B, between Bullialdus and Mercator.

DAVY. A 20-mile crater near the edge of the Mare Nubium, between Alphonsus and the Fra Mauro group. The wall is quite high in places, but in the south-east it is broken by a small, deep crater Davy A. To the south is the low-walled 14-mile crater **Lassell**, and between Lassell and Alpetragius there is a very symmetrical bright crater, Lassell B, 6 miles in diameter. Some way north-west of Davy there is a low-walled, rather incomplete ring, **Palisa**; the area between it and Davy seems to represent the ruins of a very battered and reduced walled plain considerably larger than Davy itself.

FLAMMARION. An irregular enclosure north-west of Ptolemæus and Herschel, with a maximum diameter of 45 miles. Its walls are incomplete, particularly in the north, where they have been reduced by material from the adjacent Mare surface.

FRA MAURO. This is the largest member of a group of craters in the Mare Nubium; the others are Bonpland, Parry and Guericke. Fra Mauro itself, with a diameter of 50 miles, is the largest; its walls have been reduced, and it almost, though not quite, comes into the category of a ghost. Bonpland (36 miles across) is in a slightly better state of preservation, and Parry (26 miles) has still higher walls. These three formations have common boundaries, and are crossed by various rills. Further south is **Guericke** (sometimes, though wrongly, spelled Gueriké) with incomplete walls, very broken in the north and almost levelled in places. The interior contains considerable detail, including a crater-chain, pits and ridges, several delicate rills, and one distinct craterlet, D. Apollo 14, with Astronauts Shepard and Mitchell, came down in the Fra Mauro area.

HERSCHEL. A fine crater 28 miles across, with terraced walls and a large central peak. It lies closely north of Ptolemæus. Adjoining it to the north is **Spörer**, much less complete, and showing some evidence of having been partly filled with lava. To the west of Herschel lies **Glydén**, low-walled but reasonably regular, and in the triangle formed by Herschel, Spörer and Glydén there is a splendid example of a crater-valley, easy to see with any small telescope when suitably lit.

LALANDE. A well-formed crater 15 miles in diameter, with a low central hill. Between it and Flammarion there is a very old incomplete ring, on the south wall of which is a distinct crater, D, not much smaller than Lalande itself, and of the same type.

LANDSBERG. The more correct spelling is 'Lansberg', but the alternative form has been in use for so many years that to alter it now would be pedantic. The crater lies on the Mare Nubium, north of the Riphæan Mountains, and is a fine example of a ringed plain, 28 miles in diameter, with massive walls and a central mountain. Well to the west lies the shallower and less regular **Gambart**; between Gambart and Landsberg is a distinct little crater, Gambart A; and between Gambart and Lalande there are two craterlets, **Turner** and Gambart F.

LUBINIEZKY. A very reduced crater on the Mare Nubium, north-west of Bullialdus; its walls are everywhere low, and in places discontinuous, while the floor is as dark as the surrounding Mare. To the north lies the distinct little crater **Darney**. There are numerous ghost rings in this area; **Gould** and **Opelt** are others.

MEDII, SINUS. The Central Bay. Part of it is shown in this Section; Mösting, Sömmering and Schröter lie near its boundary. On the Sinus itself are two distinct craterlets, **Bruce** and **Blagg**.

MÖSTING. A well-formed crater 16 miles across, with a low central hill. To the south-south-east is Mösting A, which is very bright and is a minor ray-centre; it has been used as a reference point for various lists of the positions of lunar features. It actually lies on the wall of Flammarion.

NICOLLET. A distinct 10-mile crater on the Mare Nubium, roughly west of Birt. Some distance away is **Max Wolf**, low-walled and irregular in outline.

NUBIUM, MARE. One of the largest of the lunar seas; much of it lies in this Section, though parts extend into Sections 9 and 11. It is lighter, patchier and much less regular than the Mare Imbrium, and there are no really high ranges to mark its border. Craters on it include Bullialdus and the Fra Mauro group. Various space-probes have landed in the Mare Nubium; the region where Apollo 12 came down is often called the 'Mare Cognitum' or Known Sea, though there really seems no justification for a separate name.

PARRY. This crater is in the Fra Mauro group, and has been described with it.

PTOLEMÆUS. One of the most famous walled plains on the Moon. It is over 90 miles across, and is near the centre of the disk, so that it is ideally placed for observation. Its floor is darkish in hue, but contains various objects as well as a well-marked craterlet, Ptolemæus A (now named Ammonius). The walls of Ptolemæus are fairly continuous, and contain some high peaks, rising to 9,000 feet or so. Ptolemæus is a member of a great chain of walled formations; the other members are Alphonsus and Arzachel.

RÉAUMUR. A low-walled ring near the edge of the Sinus Medii. Its neighbour **Oppolzer** is similar. Rhæticus (Section 13) lies south-west.

RIPHÆAN MOUNTAINS. A low mountain range on the Mare Nubium, lying along the wall of a large ghost-crater. Its highest peaks rise to no more than 3,000 feet. The **Ural Mountains**, to the north, are really part of the Riphæans. The best means of identification is provided by Euclides (Section 9), whose bright nimbus makes it easy to find under any conditions of illumination.

SCHRÖTER. A semi-ruined, 20-mile crater near the common border of the Mare Nubium and the Sinus Medii. It has been badly reduced by Mare lava, and its walls are incomplete in the south. Note that it is nowhere near Schröter's Valley, which lies in the Second Quadrant near Herodotus.

SÖMMERING. Another reduced ring, 17 miles across, close to Schröter and Mösting. Its walls are low everywhere, with a gap in the south.

STRAIGHT WALL. As has been made clear in the text, this formation is not straight, and is not a wall. It is a particularly interesting fault, between Thebit and Birt, ending to the south in a group of hills often called the **Stag's-Horn Mountains**; these peaks make up part of the boundary of an ancient ring lying to the west. Before full moon, the Wall appears as a dark line; for some time around full it is invisible, and after full it reappears as a bright line, as the Sun shines upon its inclined face.

THEBIT. A very significant crater; it is broken by a smaller formation, A, which is in turn broken by the still smaller D. It is always identifiable, even when the Moon is full. It may be a true member of the Walter chain, since its outer slopes merge with those of Purbach to the south-west. Some way north-west of Thebit lies the cape known as **Promontorium Ænarium** – a misspelling of the proper name of Tænarium, but which has become generally accepted.

TURNER. This small crater has been described with Landsberg and Gambart.

Fourth (South-East) Quadrant

This quadrant contains few seas; the Mare areas are limited to part of the Mare Fœcunditatis, all of the Mare Nectaris, a very small portion of the Sinus Medii, the irregular Mare Australe on the south-east limb, and a very slight area of the Mare Tranquillitatis – though an important one, since it was here that the Lunar Module of Apollo 11 brought Neil Armstrong and Buzz Aldrin down for that first landing in July 1969.

Most of the quadrant is occupied by rugged uplands, and there are craters and walled plains of all sorts, ranging from the huge, ruined Janssen to superb formations such as Theophilus; there are smaller rings, and almost countless craterlets. Lofty mountain ranges are absent, but there is the interesting Altai Scarp running north-westward from Piccolomini.

Very few transient phenomena have been seen in this quadrant, and the paucity of events seems to be due to something more fundamental than mere observational selection.

Apollo 11 is not the only probe to have landed in this area. Apollo 16 came down in the region of Descartes, while Russia's Luna 16 brought home rock samples from the Mare Fœcunditatis.

Section 13

ABENEZRA. This crater forms a notable pair with Azophi. The two are of about the same size (27 miles in diameter) and have high walls; the inner ramparts are terraced, particularly with Abenezra. To the south-west, just on Section 14 but shown here also, is **Playfair**, equal in size to Abenezra and Azophi, and there is a fairly distinct craterlet, A, between Playfair and Azophi. The most interesting feature of the whole group is that Abenezra overlaps a shallow formation, C, which is if anything fractionally the smaller of the pair, so that we may have a possible departure from the general rule even though the difference in size between the two is negligible.

ABULFEDA. Another interesting pair is made up of Abulfeda and Almanon, not far from Abenezra; the intervening space contains Geber. Abulfeda is larger and deeper than Almanon, with a diameter of 40 miles and walls rising to 10,000 feet; the figures for Almanon are 30 miles and 6,000 feet. A crater-valley runs from north-west to south-east between them, and extends toward the general direction of Polybius; undoubtedly it is associated with the Altai Scarp. To the north of Abulfeda are **Descartes** and **Åndel**, which have low walls, as well as the bright 6-mile **Dollond**. Dollond borders a large, ruined enclosure, and some maps give the name to the ruin instead of the deep craterlet. It was in the Descartes area that Apollo 16 landed.

ALBATEGNIUS. A great crater 80 miles in diameter, near the Ptolemæus chain (Section 12) and forming a notable pair with Hipparchus. The walls of Albategnius are generally quite high, with one or two peaks rising to well over 10,000 feet, and there are prominent terraces. The walls are, however, broken in the south-west by a large crater, **Klein**. There is a central mountain on the floor of Albategnius, as well as various craterlets; it is a fine sight under oblique lighting.

ALFRAGANUS. A very conspicuous crater. It is only 12 miles in diameter, but is very bright and as it is also the centre of a minor ray-system it is easily found at full moon; it lies in the uplands, well north-west of Theophilus and south-east of Delambre. Near it are the triangular, 30-mile **Hypatia**, almost on the Mare border; the elliptical **Taylor**, 25 miles across, with a group of three craters to its north-west; and the low-walled, rather irregular **Zöllner**.

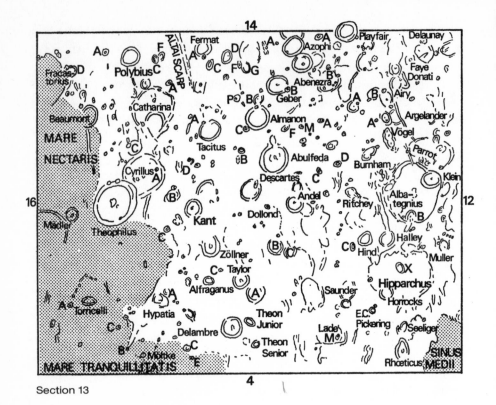

Section 13

ALMANON. This crater has been described with its companion, Abulfeda.

ALTAI SCARP. This is certainly more of a scarp than a mountain range, and is part of the ring-system of the Mare Nectaris; most of the Mare lies in Section 16. The Scarp rises to an average of 6,000 feet above the general level to the east, but only very slightly above the level to the west. Most of it lies in Section 14, but it extends past Fermat almost as far as Tacitus. **Fermat** itself is 25 miles in diameter.

ARGELANDER. A 20-mile crater south-east of Albategnius, with terraced walls and a central peak. Its twin is the nearby **Airy**, rather similar to it, and also with a central peak. To the east, a crater-rill runs across into Parrot.

AZOPHI. This adjoins Abenezra, and has been described with it.

BEAUMONT. An excellent example of a bay. It lies on the edge of the Mare Nectaris, between Fracastorius and Cyrillus, and is 30 miles in diameter; it retains its landward wall, but the rampart on the Mare side has been largely destroyed by lava. There are some very small, delicate craters on the floor, but little else.

BURNHAM. A low-walled formation south-east of Albategnius.

CATHARINA. The southernmost member of the Theophilus chain. It is about 55 miles in diameter, with rugged walls; the floor contains a large, low-walled, ruined ring in the north. There is no central peak, so that, as with other trios, we have a gradation in type from Theophilus through Cyrillus to Catharina. The area between Catharina and Cyrillus is high, and is a splendid sight under oblique illumination.

CYRILLUS. Cyrillus lies between Catharina, from which it is separated by the upland area described above, and Theophilus, which overlaps it. As usual, the walls of the broken formation (Cyrillus) remain perfect up to the point of junction, ruling out any violent method of origin. The floor of Cyrillus contains a reduced central hill, as well as a considerable crater, Cyrillus A, and much fine detail. The diameter is about 60 miles. Between Cyrillus and Kant, to the north-west of Cyrillus itself, is a distinct crater, B.

DELAMBRE. A well-formed 32-mile crater west of the Mare area. It has high walls, with some peaks reaching to 15,000 feet. This was the landing area of the unmanned probe Ranger 8, which obtained excellent photographs of Delambre during its plunge Moonward. To the west lie the two Theons, both of which are bright. Delambre contains a central peak with a summit pit.

DELAUNAY. A most peculiar compound formation to the north of Blanchinus, north-east of Purbach (Section 11); Purbach, Lacaille, Delaunay, Faye and Donati are roughly lined up. Delaunay is made up of two irregular structures with a ridge between them, common to both.

FAYE. This and **Donati** are two irregular, imperfect formations near Delaunay. Each is about 22 miles in diameter, and each has a central peak.

FRACASTORIUS. This great 60-mile bay lies partly in this Section and partly in Section 16; it marks the southernmost point of the Mare Nectaris. The seaward wall has been almost destroyed, but is still traceable. There is a darkish streak across the floor which is of a slightly reddish hue, and is detectable with a moonblink device, making a good test for atmospheric conditions of observation. Abutting on it to the west is an irregular, somewhat triangular depression, D.

GEBER. A regular, 25-mile crater between Almanon and Abenezra. Its walls are high and terraced, but disturbed to the west by a much smaller crater, Geber B.

HALLEY. This and **Hind** are two well-formed terraced craters close to Hipparchus and Albategnius. Halley is 22 miles in diameter, Hind 16. A valley runs from Halley into Hipparchus in the one direction, and to the eastern glacis of Albategnius in the other. West of Halley, between Hipparchus and Ptolemæus, is the low-walled, irregular formation **Müller**, and some way south of Hind is another deformed object, **Ritchey**.

HIND. This crater has been described with its larger companion, Halley.

HIPPARCHUS. This tremendous enclosure, almost equal to Ptolemæus in size, has been very broken but is still striking when seen under low illumination.

A grid system is very evident in this area, and Hipparchus contains a great amount of fine detail as well as the prominent 18-mile crater **Horrocks**. Outside Hipparchus, to the north-east, are the reduced formations **Saunder** and **Lade**, and the more regular but smaller crater **E. C. Pickering**. Hipparchus is, of course, much less imposing than Ptolemæus (Section 12).

KANT. A very deep crater, 20 miles in diameter, west of Theophilus. Notable for its huge, rounded central mountain, which is crowned by a summit pit. Kant is one of the few craters in this quadrant in which a transient event has been reliably reported – by the French astronomer Trouvelot in January 1873; on the 4th of that month he stated that the crater was 'filled with mist'.

MÄDLER. A prominent irregular crater 20 miles across, on the Mare Nectaris east of Theophilus. Its floor is crossed by a ridge, which joins the central mountain mass. Mädler lies near the western border of a ghost-ring which extends in the direction of Isidorus (Section 16).

MEDII, SINUS. A small part of the Sinus is shown in this Section, in the area of Rhæticus, Theophilus and the bays of Beaumont and Fracastorius.

PARROT. A very complex, irregular and compound structure, around 40 miles across, south of Albategnius. A crater-valley runs into it from the area of Argelander and Airy.

POLYBIUS. A fairly regular 20-mile crater south-west of Catharina.

RHÆTICUS. This and **Réaumur** (Section 12), each 28 miles in diameter, lie on the border region of the Sinus Medii. Both are somewhat reduced, with low walls; **Seeliger**, between Rhæticus and Hipparchus, is even more distorted.

TACITUS. A somewhat polygonal formation, 25 miles in diameter, between Catharina and Abulfeda. The walls are terraced, and rise in places to 11,000 feet. The floor contains two craterlets, one of which is almost central.

THEON JUNIOR and THEON SENIOR. Two prominent craterlets, respectively 10 and 11 miles in diameter, west of Delambre. Theon Junior is slightly the more brilliant of the two, but both are bright. In many ways they resemble Alfraganus.

THEOPHILUS. There can be little doubt that Theophilus is, with the possible exception of Copernicus, the grandest crater on the whole Moon. It is extremely deep, with walls rising to 18,000 feet above the floor; there is a splendid, many-peaked central mountain mass, and the inner ramparts are terraced. It is always a magnificent sight, and may be recognized under any conditions of illumination. It is the northern member of the chain of three great walled plains, and actually intrudes into its neighbour Cyrillus.

TORRICELLI. A curious, compound formation on the Mare, north of Theophilus. It is made up of two rings; the larger one, to the east, has a diameter of 12 miles. The general impression is that of a pear-shaped enclosure.

There are interesting features inside Torricelli, including some delicate rills. The small crater Torricelli B shows strange variations in brightness – due probably to TLP activity.

TRANQUILLITATIS, MARE. A small part of this large Mare lies here; the rest lies in Section 4. Near the border is a small, distinct crater, **Möltke**. This is, of course, the region in which Apollo 11 came down.

VOGEL. A peculiar formation near Albategnius, north of Argelander. It consists of three craters which have merged, so that it may be classed as a short crater-chain. There are various valleys in the neighbourhood. North-east of Vogel lies the obscure crater **Burnham**.

Section 14

ALIACENSIS. A noble crater 52 miles in diameter. It lies outside the wall of Walter, and has a rather smaller 'twin', Werner. Aliacensis has broad, terraced walls and a central mountain. South of it lie two somewhat broken rings, **Kaiser** and **Nonius**.

ALTAI SCARP. Much of the Scarp lies in this Section; it begins near Piccolomini, and runs north-westward to Tacitus in Section 13.

APIANUS. A 39-mile crater with high terraced walls, rising in places to 9,000 feet. It lies east of the Aliacensis-Werner pair, and in between it and Werner is a large, broken enclosure which has been called **Krusenstern**. South-east of Apianus is the very irregular **Poisson**, which is a compound structure with a mean diameter of about 45 miles.

BIELA. A 46-mile crater rather near the limb, not far from Pontécoulant and the Vlacq group. It has high, terraced walls, and a central peak. To the north-east the wall is disturbed by the intrusion of a considerable crater, Biela A.

BLANCHINUS. A crater north-west of Werner; it is 33 miles in diameter. Its walls are high in places, but are somewhat uneven, and the floor contains much fine detail. Craters near it include Faye and Delaunay (Section 13).

BOGUSLAWSKY. A major formation, 60 miles across and with high walls containing peaks up to 11,000 feet. It is too near the limb to be well seen.

BOUSSINGAULT. Another great formation, along the limb eastward from Boguslawsky. It is interesting inasmuch as it consists of three rings, with a maximum diameter of 70 miles. Even worse placed, and seen only under favourable libration, are **Helmholtz** (60 miles in diameter) and **Neumayer** (50 miles). In 1954 I detected some curious ray-like features crossing Helmholtz, not like any others known to me; and even after studying Orbiter and Apollo photographs I am still uncertain as to their nature.

BUCH. A regular crater, 30 miles across, north-east of Maurolycus. The floor is relatively smooth, though it includes some low-rimmed pits. Adjoining it

Section 14

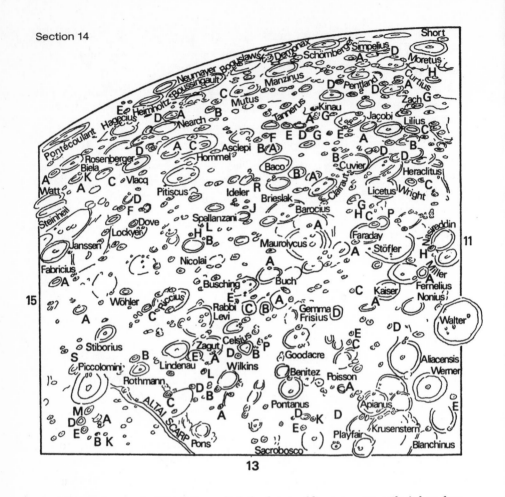

to the north east is **Büsching**, slightly larger (diameter 36 miles) but less regular. The region surrounding Buch and Büsching is rich in craterlets.

CURTIUS. A deep 50-mile crater in the south polar region, not far from Moretus, with massive, terraced walls.

CUVIER. A moderately regular, 50-mile crater, with high terraced walls and a central hill. It lies in the uplands, east of the Heraclitus-Licetus group. Close by are **Clairaut**, which has been deformed by the intrusion of two distinct craterlets, and **Baco**, 40 miles across, which has lofty walls and a low central peak. North of Baco lies **Breislak**, which is slightly smaller; between the two is a small deep craterlet. There are two well-formed craters of some size between Baco and Clairaut.

DEMONAX. This great enclosure, 75 miles in diameter, lies between Boguslawsky and the limb, but is visible only under favourable libration. This is a pity, since it is majestic and complex.

FABRICIUS. A walled plain 55 miles in diameter, breaking into the vast ruin Janssen. There is a central mountain on the generally rough floor, and the walls attain 9,500 feet in places. A crater-valley links it with the crater A, to the north, while a long rill runs northward from outside the west wall of Fabricius. To the north-east is Metius (Section 15) which is of about the same size as Fabricius.

FARADAY. A very irregular formation, around 40 miles in diameter, which has broken into Stöfler. There are two deep craters which break into the walls of Faraday itself, and the whole terrain is very complex. The floor of Faraday is much rougher than that of Stöfler.

GEMMA FRISIUS. A large, irregular enclosure north of Maurolycus, much broken by smaller craters, and with a decidedly rough floor. In the north it is disturbed by the rather more regular, 30-mile **Goodacre**, which has a low central peak.

HAGECIUS. A most peculiar formation; it is a member of the Vlacq group, south-west of Janssen. Its diameter is about 50 miles, but one wall has been ruined by the intrusion of no less than five craters. The floor is relatively smooth, but contains several craterlets.

HERACLITUS. A strange, irregular enclosure, with a central ridge. It adjoins Licetus, and Cuvier may also be regarded as a member of the group. Licetus is 46 miles across, with uneven walls and a low central hill; the rampart has been broken on the south, so that the interior connects with that of Heraclitus, though Cuvier remains separate. The whole group can be very prominent when well placed, and is not difficult to find, since it lies not far south of the dark-floored and always prominent Stöfler.

HOMMEL. A large walled plain, 75 miles in diameter, sufficiently far from the limb to be quite well seen under good libration. It has walls of fair height, and is notable because the floor has been broken by two large craters, A and C; the eastern (A) has a central mountain. Outside Hommel, to the west, are the 20-mile craters **Asclepi** and **Tannerus**, while Pitiscus adjoins Hommel to the north.

JACOBI. A 41-mile crater, with walls rising in places to almost 10,000 feet. It lies south-east of the Heraclitus group. Its neighbours are **Kinau** (26 miles across) and **Lilius** (32 miles), both of which have high walls and central peaks, but present no features of particular note. A considerable crater lies between Lilius and Jacobi.

JANSSEN. This great ruin has a diameter of over 100 miles, and is thus one of the Moon's major formations, but it is in a sad state, and is prominent only when near the terminator. Its walls are broken in the north by Fabricius and in the south by the bright-walled, 30-mile crater **Lockyer**; the floor is light and is crowded with detail, including a prominent crater-rill.

KINAU. This well-marked crater has been described with Jacobi.

LICETUS. A member of the Heraclitus group, and described with it.

LILIUS. This has already been described with Jacobi.

LINDENAU. This is described with Rabbi Levi.

MANZINUS. A large crater, 55 miles across, and with high terraced walls reaching 14,000 feet here and there. It lies in the Boguslawsky area, rather further away from the limb.

MAUROLYCUS. A walled plain with an average diameter of 68 miles, lying east of the darker-floored Stöfler and fairly obviously associated with it. The walls are of some altitude, but are broken in several places by craterlets and landings. There is a central mountain group, and the floor in general is rough, containing some ruined rings. Immediately outside Maurolycus, to the south-east, is **Barocius**, 50 miles across, whose high walls are broken in the north-east by two considerable craters. One of the most interesting points about the Maurolycus-Barocius group is that Maurolycus itself has encroached upon an old formation to its south; the broken formation – which also touches Barocius – is smaller than Maurolycus, so that here we do seem to have an exception to the general overlapping rule.

MILLER. This and **Nasireddin** are two 30-mile craters west of Stöfler. They make up part of the group which includes Orontius and Huggins, and which has been described in Section 11.

MORETUS. A fine walled plain 75 miles in diameter, and comparable with Theophilus and Copernicus; it is indeed larger than either, and has a high, broad, terraced wall. The floor is somewhat dark, and there is a particularly lofty central mountain crowned by a small pit. Nearby is the 29-mile, high-walled crater **Cysatus** (Section 1). The mountains on the limb beyond Moretus were once thought to be the highest on the whole Moon, and were named the Leibnitz Mountains; but space-probe pictures have shown that they do not make up a true range, and the name has now been deleted from the official maps, though it has been reallotted to a crater on the Moon's far side.

MUTUS. A 50-mile crater near Manzinus. Its walls are high, rising to peaks of 14,000 feet. Mutus has two large craters on its floor, so that it is easy to recognize; the arrangement resembles that of the distinctly larger Hommel. The area between Mutus and Tannerus, to the north, is very crowded with detail.

NASIREDDIN. This has been described with its companion Miller.

NEARCH. This 38-mile formation lies near Hommel and Hagecius, and may be regarded as a member of the Vlacq group. The floor contains several craterlets of some size, and immediately to the south is a deep crater, Nearch A, which has a central peak.

NICOLAI. A regular crater 27 miles in diameter, roughly between Janssen and Maurolycus. Nicolai has walls rising to 6,000 feet. It is right in the upland area, but there are no large craters really close to it, and this is one of the 'smoother' parts of the area, even though it is still very rough judged by general lunar standards. Well to the south-west lie various craters of little note, including **Spallanzani** and **Ideler**.

PENTLAND. A 45-mile crater near Curtius, with terraced walls and a double-peaked central mountain. A considerable crater, Pentland A, lies to the south. Nearby lies **Zach**, which is about the same in form.

PICCOLOMINI. A splendid crater at the end of the Altai Scarp. It is 56 miles across, with high, terraced walls which include peaks rising to 15,000 feet; the southern wall has unusual structure. To the north-west, on the eastern side of the Scarp, there is a dome. Between Piccolomini and the Lindenau group lies the well-formed crater **Rothmann**, which has a peak slightly displaced from the centre of the floor.

PITISCUS. A crater 50 miles in diameter, adjoining Hommel to the north. Its walls have peaks rising to 10,000 feet, but are narrow in places; the floor contains several craters. To the west, in the general direction of Maurolycus, lie several craters, including Ideler and Spallanzani.

PONS. A 20-mile crater close to the Altai Scarp, north-west from Piccolomini. Its walls are abnormally thick. Nearby, to the west, is the irregular, 52-mile **Sacrobosco**.

PONTANUS. A well-formed 28-mile crater, about midway between the Aliacensis-Werner pair and the Altai Scarp. Its broad walls are disturbed in places. There is no central peak, but a crater lies very near the centre of the floor. Closely south-east is a smaller crater, Pontanus C.

PONTÉCOULANT. This lies on the borders of the present Section and Section 15. It is too near the limb to be well seen, but it is a major walled plain with a diameter of 60 miles, and there is considerable detail inside it.

RABBI LEVI. One of a group lying some way south-west of Piccolomini; the other members are **Riccius, Lindenau, Zagut, Celsius** and **Wilkins**. Rabbi Levi is 50 miles across, but is irregular in shape, with rather low walls and a rough floor containing numerous pits and craterlets. Celsius, separated from Rabbi Levi by a broad valley, is smaller, and has a deep craterlet some way from the centre of the floor. Riccius (50 miles in diameter) also has broken walls and a rough, pitted floor; Lindenau is 35 miles across, with higher terraced walls and a group of low mounds near its centre; Wilkins, an irregular enclosure with a mean diameter of 40 miles; Zagut, another formation of the same type, though larger (50 miles across). The whole group is complex, but in no way particularly notable.

RICCIUS. This has been described with Rabbi Levi.

ROSENBERGER. This is included in the Vlacq group. It is 50 miles in diameter, with a darkish floor and a low central peak; in the south its wall adjoins that of a smaller crater.

SCHÖMBERGER. A large crater, over 40 miles in diameter, westward along the limb from Boguslawsky. Still closer to the limb, and very hard to study from Earth, are **Scott** and **Amundsen**, which cannot be shown in a mean libration map; Amundsen, the shallower, seems to have been damaged by the wall of Scott. Rather better placed, again not far from Schömberger, is the rather distorted but quite prominent **Simpelius**.

STEINHEIL. This and **Watt** form another splendid example of a pair of overlapping craters; they lie outside the south-west wall of Janssen. Steinheil is 45 miles across, with walls which rise to 11,000 feet in the west; Watt has some ridges and delicate craterlets on its floor. The pair is similar to that of Steinheil, but is considerably larger.

STIBORIUS. A well-marked 23-mile crater, south of Piccolomini; it has broken into a larger, older ring. The floor contains a central peak – actually slightly asymmetrically-placed. Further south lies the smaller, rather elliptical **Wöhler**, which has no points of special interest.

STÖFLER. A grand enclosure, 90 miles across, with a darkish floor which makes it identifiable under any conditions of lighting. Part of the rampart has been destroyed by the intrusion of Faraday. To the north is the 40-mile, rather irregular **Fernelius**.

VLACQ. A deep, well-formed crater 56 miles in diameter, with a central hill and walls rising to 10,000 feet in places. It is a member of the group which includes Hommel, Nearch, Rosenberger, Pitiscus and Hagecius.

WERNER. The 'twin' of Aliacensis, east of the Walter chain. Werner has a diameter of 45 miles, and is extremely regular, with high terraced walls with peaks attaining almost 15,00 feet. There is also a splendid central peak.

WILKINS. One of the Rabbi Levi group, and described with it.

WRIGHT. A well-formed 18-mile crater some way west of Licetus.

ZAGUT. This also has been described with Rabbi Levi.

Section 15

AUSTRALE, MARE. The so-called Southern Sea could not be fully explored before the space-probe era, because it lies so close to the limb as seen from Earth. It extends on to the far hemisphere, and probe pictures show that it contains various craters which are lava-flooded, though not connected – a proof that the flooding came from below (though, of course, this is now no longer a matter for debate). Mare Australe is not a well-formed sea of the

Crisium type, but it can be quite prominent even in spite of its unfavourable position. Further on the disk, between the Mare and the Rheita area, are various walled plains such as **Brisbane, Reimarus, Peirescius** and Vega. Naturally, the Mare Australe is best seen immediately after full moon, when it is on the terminator.

BORDA. A low-walled, 26-mile crater between Santbech (Section 16) and Reichenbach, west of Petavius. The interior contains some detail.

BRENNER. A very irregular formation abutting on Metius and Janssen.

FRAUNHOFER. A crater 30 miles in diameter, east of the Rheita Valley and south of Furnerius. Two craters break the north-west wall.

FURNERIUS. A great walled plain 80 miles across. It is a member of the great 'Eastern Chain' which includes Petavius, Vendelinus, Langrenus, the Mare Crisium and Endymion. The walls of Furnerius are somewhat broken, particularly in the north; the floor contains details such as craterlets, hills and delicate rills, as well as one prominent bright crater, Furnerius B.

HANNO. A dark-floored, 40-mile crater near the limb in the region of Pontécoulant (Section 14).

HASE. An enclosure 48 miles in diameter, closely south of Petavius. The walls are broken and the shape rather irregular; the floor contains many tiny craterlets. It is disturbed in the south by a smaller but deeper crater, **Hase D.** To the outer north-west of Hase it has been said that the ridges show up at sunset as an illuminated cross, but I have yet to see this appearance, though I have often looked for it.

HUMBOLDT, WILHELM. On the limb and close to the border of Sections 15 and 16, this formation is 120 miles in diameter. Until photographed from the Orbiters its details were unknown, because of its extreme foreshortening; it proves to be an irregularly-outlined structure with a magnificent system of rills on its floor, which are hidden from Earth-based observers. The

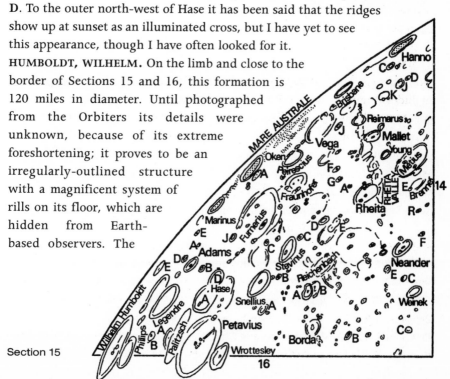

Section 15

16

westward companion of Wilhelm Humboldt is **Phillips**, 75 miles in diameter, with a long central ridge. Between Phillips and the east wall of Petavius there is a distinct crater, B.

LEGENDRE. A low-walled crater 46 miles in diameter, between Hase and Wilhelm Humboldt. The floor contains a discontinuous central ridge. South of Legendre is **Adams**, which has a central crater; closely outside it is Adams A, of considerable depth and with very bright walls.

MARINUS. A 30-mile crater, roughly between Furnerius and the limb. It has moderate walls, and a central hill.

METIUS. This large walled plain belongs to the Janssen group (Section 14). It is 50 miles in diameter, with terraced walls including one peak rising to 13,000 feet. The floor contains a considerable craterlet, B. Metius may be regarded as the twin of its south-western neighbour Fabricius, which has already been described in Section 14.

NEANDER. A well-formed crater 30 miles across, between Piccolomini and Reichenbach. The inner slopes of the wall contain two craterlets. Some distance north-westward is the smaller but quite conspicuous crater **Weinek**.

OKEN. A 50-mile crater along the limb northward from the Mare Australe, easy to find because of its dark floor. The walls tend to be linear in places, but rise here and there to 6,000 feet. To the west is Oken A, which also has a darkish floor upon which stands a low central hill.

PALITZSCH. This extraordinary formation lies closely outside the eastern wall of Petavius. It was described as an irregular gorge-like structure, 60 miles long by 20 miles wide, possibly formed by a meteorite ploughing through the lunar surface layers; but when I examined it in 1952, using the 25-inch Newall refractor then at Cambridge Observatory, I found that it is nothing more nor less than a vast crater-chain, made up of several major rings which have coalesced. Outside, to the east, is Palitzsch A, which has continuous walls and three obvious hills upon its floor.

PETAVIUS. A magnificent crater, certainly one of the finest on the Moon. It is over 100 miles in diameter, and has walls which rise to 11,000 feet; the ramparts are very complex, and in places double. The slightly convex floor contains a grand central mountain group; the main peak exceeds 5,500 feet. A particularly conspicuous rill runs from the central area to the south-west wall. Under low and moderate illumination Petavius dominates the whole area, but it becomes obscure under very high light, and is none too easy to find at full moon. It is, of course, a member of the Eastern Chain. Palitzsch lies closely east, and abutting on Petavius to the west is the well-marked 34-mile crater **Wrottesley**, which has a twin-peaked central mountain and walls rising to 8,000 feet in places.

PHILLIPS. This has been described with its larger companion, Wilhelm Humboldt.

REICHENBACH. In itself Reichenbach is not remarkable; it is rather irregular, and about 30 miles across. North of it is Reichenbach A, which has broken into a slightly larger crater to the west of it. The chief interest in this area is the splendid Reichenbach Valley, south-east of Reichenbach and narrowing steadily as it passes southward. It is really a crater-chain, and so is similar in nature to the Rheita Valley, though it is neither so conspicuous nor so well formed.

RHEITA. A crater 42 miles across, with walls rising to 14,000 feet; the crests are unusually sharp. Associated with it is the important **Rheita Valley**, which has been fully described in the text, and is easily seen with a small telescope when suitably lit. It is 115 miles long, and the breadth across the widest part is about 15 miles. In the Valley there are various unremarkable rings, such as **Mallet** and **Young**.

SNELLIUS. This and **Stevinus** form a good example of twin formations. Each is high-walled, and about 50 miles in diameter; each has a central mountain. Snellius has rather the lighter floor. Both are easy to find except under a very high light.

STEVINUS. This is described above, with Snellius.

VEGA. A well-defined crater, 50 miles across, in the Mare Australe region. Fairly deep, with more or less continuous walls, there is little detail on the floor.

WROTTESLEY. This has been described with Petavius.

Section 16

ANSGARIUS. A large ring, 50 miles in diameter, inconveniently close to the limb. The floor contains little visible detail apart from a few low hills. Slightly further on the disk, and closer to the equator, is a similar though slightly smaller crater, **La Peyrouse**, 45 miles across; there are various other lesser rings nearby.

BEHAIM. A 35-mile plain, with high walls and a central crater. The floor also includes a rill. Behaim lies between Ansgarius and Hekatæus.

BELLOT. A 12-mile crater between Crozier and Magelhæns. It is notable because its floor is exceptionally bright.

BIOT. A 10-mile crater near the edge of the Mare Fœcunditatis, between Petavius and Santbech. A rill runs from the south wall of Biot toward the west wall of Petavius, but is a rather delicate object.

BOHNENBERGER. A rather low-walled crater, 22 miles across, on the edge of the Mare Nectaris. There is a gap in the north wall, and a ridge runs across the floor. A crater of the same size, but much less well marked, lies to the south; between this and Bohnenberger, on the rim of the old ring, is a deep craterlet. There are some very delicate rills in this area.

CAPELLA. An interesting crater in the uplands just clear of the Mare Nectaris. It is about 30 miles in diameter, and has walls which are remarkably broad in view of their moderate height. The floor contains a particularly large, rounded central mountain, crowned by a craterlet. A notable crater-valley runs through Capella, and is traceable for a long way to either side. Capella has intruded upon a formation of similar size, Isidorus, lying west of it; the floor of Isidorus is relatively smooth apart from one prominent and decidedly deep craterlet.

CENSORINUS. A very small craterlet, 3 miles in diameter, in the uplands between the Mare Fœcunditatis and the southern extension of the Mare Tranquillitatis. It is one of the brightest points on the whole Moon, and is always conspicuous, particularly under high illumination. It lies on a bright patch. There is a crater, A, to the east, and a moderately large but low-walled and irregular formation to the west, touching the edge of the Mare Tranquillitatis to the north-east of Torricelli (Section 13).

COLOMBO. A 50-mile enclosure in the upland between the Mare Nectaris and the southern part of the Mare Fœcunditatis. It is disturbed in the north-west by Colombo A, about half the size and with a central hill. Well to the south-east lies **Cook**, which is 26 miles in diameter; the walls are low, but the crater is easy to recognize because of its dark floor. South-west of Cook is a deformed crater of similar size, **Monge**, and various other unremarkable craters, including **Maclure**.

CROZIER. A 15-mile crater, south-east of Colombo. It has a central hill.

FŒCUNDITATIS, MARE. Most of the Mare Fœcunditatis lies in this Section, though some of it extends on to Section 1. It is one of the less regular of the great seas, and there are not many large craters in it; the most interesting objects are probably the Messier twins. The Mare has no high mountain borders, and is connected with the Mare Tranquillitatis.

FRACASTORIUS. (Sometimes shortened to Fracastor.) The great bay on the coast of the Mare Nectaris. It lies partly in Section 13, and has been described there.

GOCLENIUS. This makes a pair with Gutenberg. It is 32 miles in diameter, with walls rising to 5,000 feet; as it lies on the edge of the Mare Fœcunditatis, it has been flooded by the Mare lava. It has a low central hill, and the floor is cut through by a rill which extends beyond the crater. There are various other rills in the area.

GUTENBERG. Gutenberg is larger than Goclenius, since its diameter is 45 miles, but it has been distorted, and its wall is broken in the north-east by a 14-mile crater, E. There is a reduced central peak on the lava-flooded floor.

HEKATÆUS. A large, rather pear-shaped walled plain close to the limb, adjoining Wilhelm Humboldt to the north; the line is continued by Behaim, Ansgarius and La Peyrouse. Hekatæus is clearly a compound formation, and there are ridges on its floor. Several lesser rings lie near it, further on the disk.

HUMBOLDT, WILHELM. This vast enclosure lies mainly in Section 15, and has already been described. Part of Phillips is also shown in the previous Section.

ISIDORUS. The companion of Capella, and described with it.

KÄSTNER. Yet another large crater too close to the limb to be well seen; it continues the chain from Wilhelm Humboldt through Hekatæus, Behaim, La Peyrouse and Ansgarius along to the Mare Smythii. Kästner is 80 miles in diameter; the floor is relatively smooth apart from a ridge running from the north wall to the centre. Between Kästner and the limb is a smaller but still considerable crater, Kästner B.

LANGRENUS. A tremendous walled plain, 85 miles across. It is a member of the eastern Chain, and is in every way compatible with Petavius. The walls are high, massive and terraced, rising to 9,000 feet, and the floor contains a bright, twin-peaked central mountain mass. Langrenus is most imposing under low light, and appears as a bright patch near full moon. To the north-

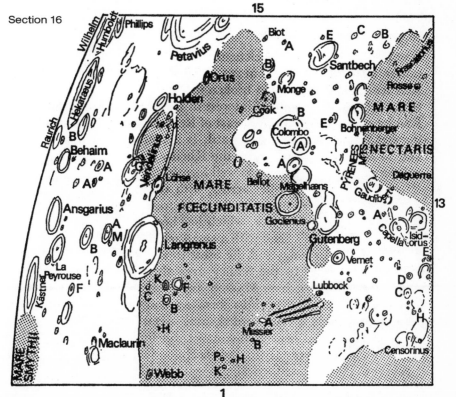

west are three craters forming a triangle: B, K and F. It was in Langrenus that Dollfus detected TLP, in 1992.

LA PEYROUSE. This crater has been described with Ansgarius.

MAGELHÆNS. A 25-mile crater south of Goclenius. It has a rather dark floor and is joined to the south-east by a smaller crater (A) with a central hill, so that Magelhæns gives the impression of being a double formation.

MESSIER. This and its companion Messier A are described in the text. The curious double 'comet' ray spreads across the Mare in the general direction of **Lubbock**, an 8-mile, fairly bright craterlet; some way to the north-west of Lubbock is an irregular formation, N, with a smaller companion. Messier A used to be called 'W. H. Pickering', but this name has been officially dropped.

MONGE. An unremarkable crater, near Cook; described with Colombo.

NECTARIS, MARE. Part lies here, and part in Section 13. Craters along its eastern border include Bohnenberger and the reasonably regular, rather dark-floored **Gaudibert**. Note the bright, deep craterlet Rosse, well on the Mare; it is 10 miles in diameter, and lies on a bright ray. Low ridges connect it with Fracastorius. Also on the Mare, west of Gaudibert, is the very low-walled **Daguerre**.

PYRENEES MOUNTAINS. Not a true range, but a collection of moderate hills, lying roughly between Gutenberg and Bohnenberger.

ROSSE. The bright craterlet on the Mare Nectaris, and described with the Mare.

SANTBECH. A 44-mile walled plain; some parts of the walls are high, and the floor is rather darkish. A depression cuts through it. Santbech lies in the uplands between the Mare Nectaris and the Mare Fœcunditatis, almost due east of Fracastorius.

VENDELINUS. A majestic irregular formation, over 100 miles from north to south. It is considerably ruined, and seems therefore to be older than its companions in the Eastern Chain, Langrenus and Petavius, which are much the same size. The floor of Vendelinus is darkish, and there is no central peak. To the north-east the rampart is broken by the intrusion of a 45-mile crater which certainly merits a separate name (on some maps it was called 'Smith'); the smaller **Lohse**, with a central peak, intrudes on the north-west; and to the south of Vendelinus lies **Holden**, 25 miles in diameter and decidedly deep. West of Holden is another crater, B, deserving of a separate name. Vendelinus and its companions make a grand picture under oblique illumination.

WEBB. A bright 14-mile crater, with a darkish floor and a central hill; it is the centre of a very short, inconspicuous system of bright rays. It lies near the border of the Mare Fœcunditatis, close to the lunar equator. In the uplands to the east is **Maclaurin**, which has a diameter of 30 miles; the walls are uneven, and the floor is distinctly concave. The **Mare Smythii** lies on the limb in this region, but most of it is in Section 1, and has been described there.

Latin and English Names of the Lunar Seas

Mare Australe	The Southern Sea
Mare Crisium	The Sea of Crises
Palus Epidemiarum	The Marsh of Epidemics
Mare Fœcunditatis	The Sea of Fertility
Mare Frigoris	The Sea of Cold
Mare Humboldtianum	Humboldt's Sea
Mare Humorum	The Sea of Humours
Mare Imbrium	The Sea of Showers
Sinus Iridum	The Bay of Rainbows
Mare Marginis	The Marginal Sea
Sinus Medii	The Central Bay
Lacus Mortis	The Lake of Death
Palus Nebularum	The Marsh of Mists
Mare Nectaris	The Sea of Nectar
Mare Nubium	The Sea of Clouds
Mare Orientale	The Eastern Sea
Oceanus Procellarum	The Ocean of Storms
Palus Putredinis	The Marsh of Decay
Sinus Roris	The Bay of Dews
Mare Serenitatis	The Sea of Serenity
Sinus Æstuum	The Bay of Heats
Mare Smythii	Smyth's Sea
Palus Somnii	The Marsh of Sleep
Lacus Somniorum	The Lake of the Dreamers
Mare Spumans	The Foaming Sea
Mare Tranquillitatis	The Sea of Tranquillity
Mare Undarum	The Sea of Waves

APPENDIX VII

Index to Formations Described in the Map and Text

(Roman = general text; *italic* = entry in Appendix VI; **bold** = Map Sec. No.)

APPENDIX VIII
The Far Side of the Moon

To make this book reasonably complete, it seems only right to include a map of the far side; but there is no point in giving too much detail, because there are so few people who will ever be able to see the features direct! For the same reason, I have followed the official practice of putting north at the top.

There are obvious differences between the far and the near hemispheres. Apart from the Mare Orientale, which spans both, the far side has no major maria, but there are huge 'unfilled' basins such as Hertzsprung and Korolev. The general aspect is of upland, and the laws of distribution apply. There is also a certain uniformity in the distribution of the formations; I have already mentioned the great arc which includes Birkhoff, D'Alembert (not to be confused with the D'Alembert Mountains on the near hemisphere – a name not included in the latest IAU list), Campbell, the Mare Moscoviense, Mendeléev, Gagarin and Mare Ingenii, as well as various others. Among special features mention should be made of the tremendous valleys of Schrödinger and Planck, in the north, not very far beyond the Earth-turned limb at maximum libration. This, then, is a mere outline chart, and I have named only the leading formations. The nomenclature follows that approved by the IAU. The names range from those of famous novelists (Jules Verne, H. G. Wells) through to modern astronauts.

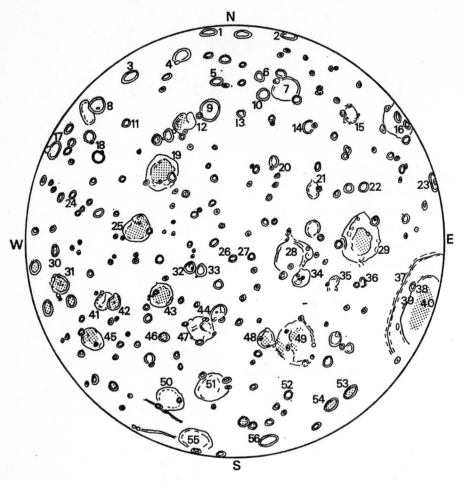

Map of the Far Side of the Moon

1 Nansen
2 Brianchon
3 Compton
4 Schwarzschild
5 Avogadro
6 Sommerfeld
7 Birkhoff
8 Fabry
9 D'Alembert
10 Rowland
11 Wells
12 Campbell
13 Duner
14 Fowler
15 Landau

16 Lorentz
17 Joliot
18 Szilard
19 MARE
  MOSCOVIENSE
20 Cockcroft
21 Mach
22 Fersman
23 Einstein
24 Fleming
25 Mendeléev
26 Dædalus
27 Icarus
28 Korolev
29 Hertzsprung

30 Hirayama
31 Pasteur
32 Keeler
33 Heaviside
34 Galois
35 Paschen
36 Ioffe
37 CORDILLERA
  MOUNTAINS
38 Lowell
39 ROOK
  MOUNTAINS
40 MARE
  ORIENTALE
41 Fermi

42 Tsiolkovskii
43 Gagarin
44 Van de Graaff
45 Milne
46 Jules Verne
47 MARE INGENII
48 Oppenheimer
49 Apollo
50 Planck
51 Poincaré
52 Minkowski
53 Mendel
54 Boltzmann
55 Schrödinger
56 Zeeman

INDEX

Callisto 18
Carpenter, James 89–90
Cassini, Giovanni 59
Cassini probe 19, 162
Ceres 17
Cernan, Eugene 131, 134–135
Charbonneaux 104
Charon 19
Chicxulub meteorite 92
Clementine probe 82, 137–138, 162
Collins, Michael 6, 107, 131
colongitude, lunar 154
Columbus, Christopher 115
comets 19–20
Conrad, Charles 127, 133
continental drift 26
Copernican Revolution 13
Copernicus 13, 34–35
core, lunar 103–104
Crab Nebula, occultation of 97
Cragg, T. A. 104
craterlets 83
craters, lunar
  atomic bomb theory of origin 88
  chains of 76, 77, 83
  classification of 78
  depth/diameter ratios of
    78–79, 81–82
  description of individual 166–222
  distribution of 81, 83
  forms of 78, 79–80, 81–83
  ice theory of origin 86–87
  impact theory of origin 92–94
  new, possibility of? 111
  origin of 86–94
  polar 137
  relative ages of 80, 84
  slope angles of 79
  tidal theory of origin 88–89
  volcanic theory of origin 89–92

craters, terrestrial 21, 92
crust, lunar 102
Cudnik, Brian 99
Cyrano de Bergerac 119

Dactyl 17
Daguerreotypes 65
Danjon Scale 116
Darwin, George 25
Davis, D. R. 27
Davison, C. 47
de la Rue, Warren 65
Deimos 17, 40
Dollfus, Audouin 97–98, 108
domes, lunar 76
Douglass, A. E. 97
Draper, J. W. 65
drawings, lunar 152–154
Duke, Charles 134
Dunham, David 99
dust theory 102, 126–127

Eagle 6
Earth
  age of 24
  second satellite of? 29–31
  status of 10–11
earthquakes 47, 89
earthshine 10, 33, 57
eclipses, lunar
  causes of 113–114
  colours during 115–116
  Danjon Scale of 116
  effects on lunar surface 117
  frequency of 114
  historical 114–115
  list of (2000–2008) 157
eclipses, solar 37–38, 112
Elger, T. G. 66
Eratosthenes 12

This revised edition first published in the United Kingdom in 2001
by Cassell & Co

Text copyright © 2001 Patrick Moore
Design and layout copyright © 2001 Cassell & Co

The moral right of Patrick Moore to be identified as the author of this
work has been asserted in accordance with the Copyright, Designs and
Patents Act of 1988.

All rights reserved. No part of this publication may be reproduced in any
material form (including photocopying or storing in any medium by
electronic means and whether or not transiently or incidentally to some
other use of this publication) without the written permission of the
copyright owner, except in accordance with the provisions of the Copyright,
Designs and Patents Act 1988 or under the terms of a licence issued by the
Copyright Licensing Agency, 90 Tottenham Court Road, London W1P 9HE.
Applications for the copyright owner's written permission to reproduce any
part of this publication should be addressed to the publisher.

A CIP catalogue record for this book is available from the British Library.

ISBN 0304 354694

Designed by Design Revolution
Jacket Design by Austin Taylor
Front jacket image © Hulton Getty

Printed in the UK by CPD, Wales

Cassell & Co
Wellington House
125 Strand
London WC2R 0BB